OBSERVERS AND MACROECONOMIC SYSTEMS

Advances in Computational Economics

VOLUME 8

The titles published in this series are listed at the end of this volume.

Observers and Macroeconomic Systems

Computation of Policy Trajectories with Separate Model Based Control

by

Ric D. Herbert

University of Western Sydney - Hawkesbury,
Sydney, Australia

KLUWER ACADEMIC PUBLISHERS

BOSTON / DORDRECHT / LONDON

A C.I.P. Catalogue record for this book is available from the Library of Congress.

ISBN 0-7923-8239-0

Published by Kluwer Academic Publishers,
P.O. Box 17, 3300 AA Dordrecht, The Netherlands.

Sold and distributed in North, Central and South America
by Kluwer Academic Publishers,
101 Philip Drive, Norwell, MA 02061, U.S.A.

In all other countries, sold and distributed
by Kluwer Academic Publishers,
P.O. Box 322, 3300 AH Dordrecht, The Netherlands.

Printed on acid-free paper

Printed in the Netherlands.

DEDICATION

To my wife Jan

who bore much of the opportunity cost.

TABLE OF CONTENTS

Abstract xi

Acknowledgments xiii

List of Figures xv

List of Tables xxi

Glossary xxiii

I Linear Systems **1**

Chapter 1: Introduction **2**

1.1 Aims of This Research 2

1.2 Overview of Control Applications in Economics 4

1.3 Outline and Significance of This Book 9

Chapter 2: Observers For Linear Rational Expectations Models **15**

2.1 Introduction 15

2.2 The Currie & Levine Approach 17

2.3 The McKibbin & Sachs Approach 23

2.4 An Observer To Learn Expectations 26

2.5 An Application 36

2.6 Conclusion 58

2.7 Appendix: The Linear Quadratic Tracking Problem 60

Chapter 3: Low Order Dynamic Macroeconometric Models **64**

3.1 Introduction . . . 64

3.2 The Multiplier Accelerator Model . . . 65

3.3 IS-LM Models . . . 67

3.4 Contemporaneous Control Time Series Models . . . 88

3.5 Conclusion . . . 107

Chapter 4: Control Strategies for Attaining Ramp Output **109**

4.1 Introduction . . . 109

4.2 Preliminaries . . . 111

4.3 Integral Control . . . 114

4.4 Feedforward Control . . . 132

4.5 Linear Quadratic Output Tracking . . . 142

4.6 Conclusion . . . 147

4.7 Appendix: Discrete Linear Quadratic Output Tracking . . . 148

Chapter 5: Visualisation Techniques **152**

5.1 Introduction . . . 152

5.2 Visual Simulation Systems Development . . . 155

5.3 Observers and Tracking Ramps . . . 161

5.4 Numerical Issues . . . 167

5.5 Data Visualisation . . . 169

5.6 Conclusion . . . 171

II Nonlinear Systems **173**

Chapter 6: Optimal Control with Nonlinear Models: Recent Studies **174**

6.1 Introduction . . . 174

6.2 Linearisation Approaches . . . 175

6.3 Nonlinear Optimisation . . . 181

6.4 Separate Model Based Control . . . 189

6.5 Conclusion . . . 191

Chapter 7: A Practical Macroeconometric Model **193**

7.1 Introduction . . . 193

7.2 The Murphy Model . . . 194

7.3 Model Solution . . . 197

7.4 Visual Simulation of the Model . . . 206

7.5 Conclusion . . . 209

Chapter 8: Linear Observer Based Control with the Nonlinear Macroeconometric Model **211**

8.1 Introduction . . . 211

8.2 Simple Observers . . . 213

8.3 Expanded Observers . . . 220

8.4 Visual Simulation . . . 231

8.5 Conclusion . . . 236

Chapter 9: Nonlinear Observers **239**

9.1 Introduction . . . 239

9.2 Preliminaries . . . 241

9.3 A Nonlinear Observer . . . 242

9.4 Application to a Low Order Model . . . 248

9.5 Macroeconometric Model Application . . . 265

9.6 Conclusion . . . 272

Chapter 10: Conclusion and Further Work **274**

10.1 Synopsis . . . 274

10.2 Extensions of this Research . . . 277

Bibliography **280**

Appendix A: Software Engineering **304**

Index **307**

ABSTRACT

Observers and Macroeconomic Systems is concerned with the computational aspects of using a control theoretic approach to the analysis of dynamic macroeconomic systems. The focus is on using a separate model for the development of the control policies. In particular, it uses the observer based approach whereby the separate model learns to behave in a similar manner to the economic system through output-injection.

The work shows how this approach can be used to learn the forward looking behaviour of economic actors which is a distinguishing feature of dynamic macroeconomic models. It also shows how it can be used in conjunction with low order models to undertake policy analysis with a large practical econometric model. This overcomes some of the computational problems arising from using just the large econometric model to compute optimal policy trajectories.

The work also develops visual simulation software tools that can be used for policy analysis with dynamic macroeconomic systems.

ACKNOWLEDGMENTS

The research reported in this book was developed during a research project at the Department of Computing, Macquarie University, Sydney, Australia. The majority of the content was included in a PhD thesis titled *Learning in the Computation of Optimal Policy Trajectories with Macroeconomic Systems.*

Thanks to my PhD supervisor Rodney Bell for his help, encouragement and enthusiasm for this project.

Thanks also to Graham Madden (Macquarie University) and David Kendrick (University of Texas) for their advice, comments and encouragement.

I would also like to thank the many other academic, public sector and private sector economists who gave their time to discuss aspects of this project. These include Rob Henderson (Dresdner Australia), Craig Louis (Australian Treasury), Warwick McKibbin (Australian National University), Chris Murphy (Econtech), Larry Shepherd (Quantec), Roger Tonkin (Macquarie University) and Geoff Weir (Bankers Trust).

Finally, thanks also to the members of the control engineering community who provided helpful comments and encouragement. In particular, I would like to thank Graham Goodwin (University of Newcastle) and Neville Rees (University of New South Wales).

LIST OF FIGURES

2.1 Interest Rate Shock . . . 40

2.2 Tax Rate Shock . . . 40

2.3 Learning the Expected Exchange Rate . . . 42

2.4 Expected Exchange Rate with Different Learning Gains . . . 43

2.5 Outputs Under Optimal Control . . . 44

2.6 Controls Generated . . . 45

2.7 Output Using Optimal Control with Stable Model . . . 46

2.8 Output Using Optimal Control with Unstable Model . . . 47

2.9 Control Generated by Both Stable and Unstable Models . . . 47

2.10 Emphasising Output: Implications for Outputs and Controls . . . 49

2.11 Price Level Under Varying Weights . . . 50

2.12 GDP Under Varying Weights . . . 50

2.13 Generated Interest Rates Under Varying Weights . . . 51

2.14 Generated Taxes Under Varying Weights . . . 52

2.15 GDP Under Fixed and Time Varying Rules . . . 53

2.16 Prices Under Fixed and Time Varying Rules . . . 53

2.17 Outputs Under Optimal Learning . . . 54

3.1 Comparison of Estimation Techniques for Gross National Expenditure . . . 79

3.2 Comparison of Estimation Techniques for Prices . . . 79

3.3 Model Multipliers . . . 81

3.4 Six State Model: Comparison for Gross National Expenditure . . . 85

3.5 Six State Model: Comparison for Prices 86
3.6 Six State Model: Multipliers 87
3.7 Six State Model: Approximating Inflation 87
3.8 Output from Numerical Example 96
3.9 Output for Inflation Rate 102
3.10 Output for Unemployment Rate 103
3.11 Output for Gross National Expenditure 105
3.12 Output for Employment 106

4.1 Continuous Model: Closed Loop 122
4.2 Continuous Model: Integral Method Using Pole Placement 124
4.3 Continuous Model: Integral Method Using LQR 124
4.4 Discrete Model: Closed Loop 129
4.5 Discrete Model: Integral Method Using Pole Placement 130
4.6 Discrete Model: Integral Method Using LQR 131
4.7 Feedforward Method 136
4.8 Continuous Model: Feedforward Method Using Pole Placement 138
4.9 Continuous Model: Feedforward Method Using LQR 138
4.10 Discrete Model: Feedforward Method Using Pole Placement 140
4.11 Discrete Model: Feedforward Method Using LQR 140
4.12 Continuous Model: Linear Quadratic Tracking 144
4.13 Discrete Model: Linear Quadratic Tracking 145

5.1 Main Simulation Window for Model in State Space Form 156
5.2 Parameters Dialogue Box Window from The Simulation Menu 156
5.3 Prices Equation in Structural Form 157
5.4 Model in State Space Form 158
5.5 Learning Expectations 159

5.6 Optimal Controller and Optimal Learning 160

5.7 Tracking Ramps Simulation Environment 162

5.8 The *Economic System* Object 163

5.9 The *Control System* Object 164

5.10 The *Desired Output* Object 165

5.11 Modified *Control System* Object 166

5.12 Linear Quadratic Regulator 167

5.13 Comparison of Output from Different Solution Techniques 168

5.14 GDP Maintained Multipliers with β_2 Varying 170

5.15 GDP Maintained Multipliers with β_2 Varying. A different view. 170

5.16 GDP Multipliers for Interest Rates with β_2 and ϕ_2 Varying. 171

7.1 Baseline GDP and Public Sector Deficit 200

7.2 Baseline Inflation Rate and Exchange Rate 201

7.3 Baseline Interest Rate and Unemployment Rate 201

7.4 Baseline Imports and Exports 202

7.5 Historical GDP and Public Sector Deficit 203

7.6 Historical Inflation Rate and Exchange Rate 204

7.7 Historical Interest Rate and Exchange Rate 204

7.8 Historical Imports and Exports 205

7.9 Visual Simulation Environment 207

7.10 Extrapolated Historical GDP and Public Sector Deficit 208

8.1 Simple Observers – Model Output 213

8.2 Multiplier Accelerator Model Output - Deviations from Desired 216

8.3 Difference Equation Model Output - Deviations from Desired 217

8.4 Multiplier Accelerator Model Control - % Deviations from Base 218

8.5 Difference Equation Model Control - % Deviations from Base 218

8.6 Unemployment Rate - % Deviations from Base 219
8.7 Output in Levels 221
8.8 Output in Deviations from Desired Output 222
8.9 GNE Results 225
8.10 Experiment 2.2 % Deviations from Historical Controls 229
8.11 Exchange Rate & Interest Rate % Deviations from Base 230
8.12 Government Spending & Unemployment Rate % Deviations from Base . . . 230
8.13 Visual Simulation Environment 231
8.14 Economic System 232
8.15 Control System 233
8.16 Output Conversion and Desired Output 234
8.17 GNE: % Deviations from Desired Output 235
8.18 Employment: % Deviations from Desired Output 236

9.1 Simulation Environment for Low Order Nonlinear Model 250
9.2 Low Order Nonlinear Model 250
9.3 Nonlinear Model: Open Loop 252
9.4 Nonlinear Model: Phase Plane 252
9.5 Nonlinear Observer: Experiment 1 254
9.6 Nonlinear Observer: Experiment 2 254
9.7 Controls: Experiment 1 258
9.8 Controls: Experiment 2 258
9.9 Outputs: Expectations Shock 260
9.10 Gains and Condition Numbers: Experiment 1 261
9.11 Experiment 4: Linear & Nonlinear Observers 262
9.12 Experiment 4: Linear Observer Failing 263
9.13 Experiment 4: Time Invariant Observer Failing 264

9.14 Nonlinear Model: GDP Parameters 266

9.15 Nonlinear Model: Interest Rate Parameters 267

9.16 *Model* Object 268

9.17 Nonlinear Observer Simulation Environment 268

9.18 *Control System* Object 269

9.19 Nonlinear Observer Results: GDP 271

9.20 Nonlinear Observer Results: Interest Rates 271

LIST OF TABLES

2.1 Condition Number of Perturbed State Equation Driving Matrix 57

2.2 Output Metrics with A_p Perturbed 57

3.1 Ordinary Least Squares Estimates for National Income 76

3.2 Ordinary Least Squares Estimates for Interest Rates 77

3.3 Ordinary Least Squares Estimates for Prices 77

3.4 Seemingly Unrelated Regression Estimates for Structural Model 77

3.5 State Space Estimates for Model 78

3.6 Six State Model: OLS Estimates for National Income 82

3.7 Six State Model: OLS Estimates for Interest Rates 83

3.8 Six State Model: OLS Estimates for Prices 84

3.9 Six State Model: 3SLS Estimates for Structural Equations 84

3.10 Six State Model: State Space Estimates 85

3.11 Stationary VAR Model: OLS Estimates for Inflation 98

3.12 Stationary VAR Model: OLS Estimates for Unemployment 99

3.13 Stationary VAR Model: 3SLS Estimates for Inflation 101

3.14 Stationary VAR Model: 3SLS Estimates for Unemployment 101

3.15 Stationarity Test Statistics 104

3.16 Non-stationary VAR Model: Parameter Estimates 104

4.1 Continuous System: Eigenvalues of Closed Loop State Equation Driving Matrix125

4.2 Discrete System: Eigenvalues of Closed Loop State Equation Driving Matrix 131

4.3 Feedforward Method: Eigenvalues of Closed Loop State Equation Driving Matrix 139

4.4 LQT Method: Eigenvalues of Closed Loop State Equation Driving Matrix . 146

8.1 Experiment Set Two Results 224

8.2 Experiment Set Two Robustness Metrics 228

8.3 Summary Metrics 234

GLOSSARY

CONDITION NUMBER: for a nonsingular matrix A relative to a natural norm $||.||_p$, the condition number, $\kappa_p(A) = ||A||_p||A||_p^{-1}$.

CERTAINTY EQUIVALENCE: the stochastic result is the same as for the deterministic case.

CONTEMPORANEOUS CONTROL MODEL: a discrete time model with state equation of the form $\mathbf{x}_{t+1} = A\mathbf{x}_t + B\mathbf{u}_{t+1}$.

CONTROLS: that subset of the exogenous variables that can be altered for economic policy (policy instruments).

DETECTABILITY: a system is detectable if the unstable modes (eigenvalues) are observable, or equivalently the unobservable modes are stable.

CONTROL SOLUTION: the data given to an econometric model for solution.

ENDOGENOUS: determined by the model.

EXOGENOUS: determined outside the model.

FORWARD LOOKING EXPECTATION: expectations based on the future rather than the past.

MODEL CONSISTENT EXPECTATIONS: a method for implementing rational expectations where the expectations variables take the values predicted by the model.

OUTPUTS: that subset of the endogenous variables of particular interest (policy targets).

OUTPUT-INJECTION LEARNING: incorporating the difference between the outputs of two models so one responds the same as the other to controls.

POLE PLACEMENT: techniques for determining feedback gains. Selection of feedback gain matrices so that the eigenvalues of the closed loop system have specific values.

POLICY TARGETS: that subset of the endogenous variables of particular interest (outputs).

POLICY INSTRUMENTS: that subset of the exogenous variables that can be altered for economic policy (controls).

RAMP INPUT: an input of a polynomial of degree one, so that $r(t) = \alpha t$.

RELATIVE ERROR: the relative error for a vector $\mathbf{y}$ from another $\mathbf{y_d}$ is $\frac{||\mathbf{y}-\mathbf{yd}||_p}{||\mathbf{y_d}||_p}$ for a natural norm $||.||_p$.

RATIONAL EXPECTATIONS: on average economic actors expectations turn out to be correct, so the expected value of a variable is its mathematical expectation.

SHOCKED SOLUTION: the data from an econometric model after solution.

STATIONARY PROCESS: the mean and variance of both the variables and the noise are constant with time.

SYSTEM TYPE: the number of conditionally unstable poles (eigenvalues). For continuous-time systems, it is the number of eigenvalues equal to zero; and, for discrete-time systems, the number of eigenvalues equal to one.

TIME-INCONSISTENCY: an optimal control rule which differs depending upon the time at which the optimisation occurs.

Part I

Linear Systems

Chapter 1

INTRODUCTION

1.1 *Aims of This Research*

This research presented in this book is multidisciplinary containing aspects of computing, control theory and economics. It uses control theory to undertake economic policy analysis, and to present this in a manner that can be used as an aid in policy development by practising economists.

This work focuses upon economic policy analysis through the attainment of desired economic outputs or policy targets. The emphasis is upon the trajectory of the outputs to a specific economic position, thereby giving an indication of the effects on these and other variables over a time path to such a position. The question of concern is, given a macroeconomic system, is it possible to develop a set of control policies that achieve a desired output trajectory, and what are the implications of such policies? This differs from the more frequently used question of what is the response of a macroeconomic system to a particular change in control instruments.

A major focus of this work is in the use of a separate model for the development of control policies to be used with an economic system. The separate model can be considered as a control object, and its task is to develop policies to be implemented on the economic system. An aim of this work is to use control methodologies developed on the control object to attain desired output trajectories, and to apply these policies to the economic system.

For the investigation of macroeconomic policy, the economic system is usually a dynamic macroeconometric model. The separation between the control object and the economic system allows the economic system to be developed using best practice econometric techniques, and, similarity, the control object to be developed using best practice control techniques. This overcomes the criticism that adding control techniques to the econometric model may not make econometric sense. It also gives more freedom to make enhancements to the control object. Further, it has the advantage of flexibility, for different control objects can be used with the same economic system, and different economic systems can be used with the same control object. The different economic systems may be econometric models based on differing economic theory; and, the differing control objects may be used to compare the policy mix between the control instruments. These type of experiments can be used to give an indication of the robustness of the policies. Theoretically, the approach could also be taken to another step where the economic system is the real economy.

An important component of this approach is that the control object learns to behave like the economic system at the same time as it develops control policies for it. This learning is through output-injection into the control object. For the case when the economic system and the control model are identical and linear, this is the control theory observer methodology. An aim of this work is to extend this methodology to include the forward looking behaviour in economic actors.

A further aim is to use output-injection learning with control models that differ from the economic system for which they are developing policies. In particular, low order linear models are used as the basis of the control object for a high order nonlinear practical econometric model. In this case, the observer has to learn about the effects of the nonlinearities, difference in order, as well as being able to develop control policies for the system.

The use of output-injection learning with nonlinear economic systems raises the issue of nonlinear observers. An aim of this work is to construct nonlinear observers for use with economic systems. Again, the aim is to consider learning with low order nonlinear models, including, for example, the learning of the forward looking expectations of economic actors through output-injection. The ability of the observer to develop control policies, is also considered. A further aim of this work, is to use low order nonlinear observers with the high order practical macroeconometric model.

An aim of this work is to present the techniques developed here in such a way they might have influence in the policy analysis debate. Many policy makers and advisors on economic policy, are unfamiliar with these techniques, so an aim is to present these techniques in a manner that can be easily comprehended and implemented. The approach is to use visualisation. Tools are constructed whereby the relationships between objects can be examined visually; and, implementation through simulation can be rapidly undertaken. In this manner, the trade off between, and amongst, outputs, controls and other economic indicators can be examined.

To be able to convince practising economists of the benefits of this approach, many and varied simulations under a variety of conditions, and with a variety of economic systems, are necessary. The aim of the visual simulation tools is to allow for such simulations. An aim of this work is to construct a variety of economic systems that can be used with the techniques. These range from low order, deterministic, linear models to a large nonlinear practical macroeconometric model.

1.2 Overview of Control Applications in Economics

The use of control theory in economics started in the early 1950's with the work of Simon (1952), Tustin (1953) and Phillips (1954). The work of Simon was on production control. Tustin noticed the similarities to control of engineering processes and macroeconomic policy making; and Phillips continued this work by introducing feedback into the stabilisation of economic models. At this time the development of economic models was rudimentary, especially with regard to dynamics. The contributions of Tustin and Phillips, did not enter into main stream macroeconomics (Petit, 1990).

Independently of this, but around the same time, Frisch (1933), Tinbergen (1952) and

Theil (1958), were developing stabilisation policy theory. Tinbergen developed the overall framework for quantitative policy analysis by distinguishing between policies which are concerned with reforms to economic institutions and those that aim to achieve the institution's economic objectives. The latter policies could be examined using quantitative analysis (Chow, 1987). The emphasis of these researchers was on the theory of instruments available to policy makers and the policy targets at which these instruments were to be aimed. An important ingredient in this was an economic model. The concern was with achieving the targets with the set of instruments using the model to show the effects of the overall policy made up of each of the policy instruments. The approach was originally with static models, then comparative static models were considered and eventually, though slowly, the theory was extended to fully dynamic models (Holly and Hughes-Hallett, 1989).

The dynamic approach was to consider discrete time, and to consider a variable at different time periods to be different variables. This allowed for the use of stacking up vectors to reduce dynamic problems to essentially static ones. This technique is widely used in economics for analysing dynamic problems.

Tinbergen and Frisch received the first Nobel Prize for economics in 1969, and this together with the development of modern control theory, led to a rush of applications of control theory to macroeconomic stabilisation policy. The early studies were undertaken by Livesey (1971), Pindyck (1972) and, in Australia, by Fitzgerald, Johnston and Bayes (1973). The Pindyck (1972) study was important for it used a linear econometric model (of the USA) and used linear quadratic optimal control techniques as means of generating economic policy. This compared with the mainly theoretical work of the earlier researchers.

Following this work, control theory "swept" (Chow, 1975, p. 149) through the economics profession during the early 1970s. The concept of constrained optimisation fits well with economists, and the minimisation of a social loss (or objective) function subject to an economic model, as is the linear quadratic optimal control approach, was seen as a useful policy development tool. Major contributions were by Pindyck (1973) and Chow (1975).

The critique came towards the end of the 1970s, and was strongly incorporated in the

"Rules versus Discretion" debate. First came the Lucas (1976) criticism which argued that econometric models are not invariant to the policy analysis they are used for, and thus should not be used in the formulation of monetary, fiscal or other economic policies.

More damaging for optimal control was the criticism of Kyland and Prescott (1977), who concluded "... that there is no way control theory can be made applicable to economic planning when expectations are rational" (Kyland and Prescott, 1977, p. 619). They justify this claim by showing that optimal control policies may not be consistent over time, with the optimal control rule being dependent upon the time the optimisation is undertaken. They further argue that policy should be made difficult to change in all but emergency situations and suggest a two year ahead announcement of policy to make "... discretionary policy all but impossible" (Kyland and Prescott, 1977, p. 633).

More recently, these strong words have been modified (Petit, 1990), and it is generally felt that the conclusions were overstated (Begg, 1982). But the 'rational expectations revolution' swept through the economics profession, and the applications of control theory abated.

The underlying criticism of the early studies was that they assume that the public sector, as policy makers, have access to sophisticated knowledge and optimisation techniques; but that other economic actors, such as the private sector, as embodied in the model do not. The models were, generally, quite simple. The public sectors' knowledge included the private sectors' response to the policy it was formulating, but the private sector had no knowledge of what the public sector was doing. Further, the models ensured that the private sector responded simply, and after the public sector introduced the policy. Developing policy with the assumption of intelligent and sophisticated governments and less knowledgeable and intelligent private sectors was not considered appropriate.

The sophisticated optimising techniques of policy makers have now also been extended to other economic actors, and incorporated into economic models. The models may now include economic actors that, for example, maximise some utility function subject to equations that explain temporal consumption and production behaviour. Such models form a large part of modern macroeconomic dynamics textbooks (e.g. Turnovsky, 1995). In 1996, this led Chow, who was a major contributor to the application of control theory to economic systems, to argued that, in this manner, "optimal control is widely applied to the study of economic behaviour" (interview quoted in Pagan, 1996, p. 151).

One response to the rational expectations critique was to consider the problem as a game between economic actors (Blackburn, 1987; Chow, 1981*a*; Chow, 1987; Holly and Hughes-Hallett, 1989; Petit, 1990; Pindyck, 1977). The usual approach was to set up a linear quadratic game between the policy makers as one player, and other economic actors as the other player. This was a natural incorporation of the developing theory of games. Research in the area is continuing. Some of this work has been taken up by control engineers in the area of robust control where the controller and the process are viewed as participating in a game (e.g. Başar and Bernhard, 1991).

Another response to the rational expectations critique was to deal with the criticism more directly. It was argued (e.g. Currie and Levine, 1993*b*) that time invariant feedback rules based on the current value of the states are time consistent. Others (e.g. Westaway, 1989), also did not see it as such a significant problem. There was also a move to "enrich" (Currie, 1985) the optimal control problem by formally incorporating 'credibility' and 'reneging' on the part of policy makers (Cohen and Michel, 1988; Currie and Levine, 1993*c*; Hall, 1986; Holly and Hughes-Hallett, 1989; Levine and Smith, 1988; McKibbin, 1989). All these responses led to large areas of continuing research.

The most important contribution of the rational expectations debate was the formal inclusion of forward looking behaviour of economic actors, in the form of rational expectations, into economic models. This was part of adding to the sophistication of the private sector in economic models. This led to the search for solution methods for both linear and nonlinear models containing rational expectations (Blanchard and Kahn, 1980; Boucekkine, 1995; Fair and Taylor, 1983; McKibbin, 1987; Taylor, 1981); as well as the search for the solution of optimal control with such models (Amman and Kendrick, 1993; Christodoulakis, Levine and Gaines, 1991; Christodoulakis, Currie, Garratt, Kemball-Cook, Levine, Barrell, Ireland and Westaway, 1993; Fisher, 1992; Gaschler, 1996; Hall and Stephenson, 1989; Hall and

Stephenson, 1990*b*; Neck and Matulka, 1994). Research in this area continues.

Economic models are normally presented in a stochastic format, so the implications of the stochastic nature of models on the optimal control problem has been a large area of research (Hall and Stephenson, 1990*b*; Holly and Hughes-Hallett, 1989; Rüstem, 1989; Turnovsky, 1977*b*). The certainty equivalence principle was introduced to control engineers from economics (Simon, 1956). Kendrick (1981) introduced the concept of learning about the stochastic nature of the control problem and especially about the stochastic model parameters through the use of adaptive control techniques (Åström and Wittenmark, 1995). The concept of learning in rational expectations models, especially in regards to optimal control, is a continuing area of research (Amman and Kendrick, 1993; Currie and Hall, 1994).

Most control applications to economics have been single entity studies, with special purpose software constructed to undertake the particular study. There has been considerable research into algorithms for solving the control problem, but most applications use specific models, specific objective functions and specific software. The construction of general purpose software tools for the easy manipulation of dynamic economic models, and especially with regard to control applications to the models is not common in the economics literature. This compares strongly with the control literature. There has been some research into such software tools for control applications to economic models (Kendrick, 1987; Kendrick and Amman, 1992; Kendrick and Amman, 1995), but it has not been a large research area.

Another difference between the control literature and the economics literature is that, control engineers tend to develop continuous time models, which may then be transformed into discrete time models, whereas economists tend to build models in discrete time. Almost all major macroeconometric models are formulated in discrete time. While there has been some research into continuous time models (Bergstrom, 1976; Gandolfo, 1981), most control applications have been in discrete time. The use of continuous time models is, though, becoming more common in the economics literature (McKibbin and Sachs, 1991; Petit, 1990; Turnovsky, 1995).

Historically optimal control applications to economics have followed the path of many ideas.

With the advent of modern control theory, linear quadratic optimal control became fashionable in economics. At the same time as its popularity waned with control engineers, it became less fashionable in the economics community. Recently though control applications to the study of macroeconomic dynamics have become more popular. Interestingly, this is as the rational expectations 'revolution' is waning. In 1996, Pagan (1996, p. 151) stated "...the past few years have seen a re-surgence of work on control." Further, one of Australia's major dynamic macroeconometric model builders argues that "optimal control is the appropriate tool for designing active macroeconomic policies" (Murphy, 1996*b*, p. 159). This work contributes to the resurgence.

1.3 Outline and Significance of This Book

The partnership between control engineering and economics has been an interesting one, in that there has been a flow of ideas and techniques between the professions. Techniques and tools that are today common in one profession were originally sparked by ideas from the other. These ideas were often standard in the originating profession. This cross-fertilisation has been beneficial to both professions. This work contributes to this cross-fertilisation. In some ways, it takes standard control theory concepts and tools and applies them to economic problems. But these are not standard control theory type problems. The concepts of the observer, the analysis of dynamic models in state space form, the use of Laplace and z-transforms for linear models, and the separation of the control object from the process for which it is developing control policies, are standard control theory practices. Yet they have not been extensively used in economics. Large models with many inputs, many outputs, and where the outputs are strongly influenced by exogenous variables, are the norm in economics. But they are not common in the control theory literature. So, too, are discrete models with the control contemporaneous with the output. Further, the concept that the 'plant' is equally knowledgeable as the 'controller' and may make informed decisions as to what it expects the 'controller' to do, and even modify its own behaviour to 'control' the 'controller', is not a common control engineering concept.

This work is in two Parts; with the first Part focusing on linear systems, and, the second Part, focusing on nonlinear systems. In the first Part, it specifically examines the issue of linear quadratic optimal control with linear rational expectations models. It shows how a wide variety of economic models can be constructed using state space techniques so that they can be used with control algorithms. It also examines the problem of control techniques to track multiple ramp outputs, for such outputs are common in economics. The first Part also constructs software which aids in the development of control experiments with economic models; and also helps in the presentation of the ideas of this work.

The second Part considers nonlinear systems. It starts with a review of a number of current studies of the application of control techniques to economic models. It also considers an old study, for it, like this work, uses a separate model for the development of control policies. In this second Part, a high order (258 state), nonlinear, practical macroeconometric model of the Australian economy is constructed. The issue of nonlinear observers is also considered, and nonlinear observers are constructed. Using the high order nonlinear macroeconometric model, control experiments are undertaken using linear and nonlinear observers.

Turning to the contents of this work in more detail, Chapter 2 examines the use of linear quadratic optimal control with linear models that include rational expectations. It shows how forward looking expectations variables can be learnt using output-injection. More importantly, it shows how conventional linear quadratic optimal control techniques can be used with such models.

This Chapter uses the standard control theory application of the observer in linear models. Its significance is that it extends this theory to what has been a considerable problem in the economics literature *viz.* that conventional linear quadratic optimal control techniques cannot be used with rational expectations models because the resultant control policies are time-inconsistent.

Chapter 3 is concerned with the construction of a range of low order linear dynamic econometric models. These models are transformed into a linear state space form. This form differs from the usual form used by control engineers in that it allows for level constants and exogenous variables. It also differs from the form frequently used with economic models in that it uses an output equation. The Chapter shows how a large range of economic models can be transformed into this state space form.

The models are estimated using Australian data, both with standard econometric techniques, and through the use of an estimation procedure that uses the state space form. The latter procedure is a form of nonlinear least squares that uses in its objective function the error between the model outputs and the historical data trajectories.

Chapter 3 also focuses on contemporaneous control models. These are discrete time models where the forcing effect of the control is at the same time as the state update. In the usual control engineering discrete state space model, the control is one period behind the state update. Contemporaneous control models are common in the economics literature. For example, the national accounting identity, $Y_t = C_t + I_t + G_t$, has all variables being contemporaneous. The common vector autoregressive model also has contemporaneous control. Contemporaneous control models have created difficulties in transforming models into a state space form. The Chapter shows how such transformations can be undertaken.

This work focuses upon macroeconomic policy analysis using control techniques. The emphasis is upon desired output trajectories for multiple outputs. For many economic variables, the trend over the relevant time horizon approximates a ramp. For this reason a set of ramp outputs is useful for a desired output trajectory. Chapter 4 develops control strategies for linear models to attain a set of ramp outputs.

The Chapter develops two methods for time invariant control rules. Such rules are time consistent (Currie and Levine, 1993*b*). The first method 'integrates out' the error between the desired and actual outputs, and can be implemented by augmentation of the system with additional states. The second method uses model inversion to generate a feedforward controller that ensures the desired set of ramp outputs are attained. The Chapter also considers the linear quadratic output tracking problem for attaining a desired output of a set of ramps.

Economic models normally have uncertain parameters. The Chapter incorporates this uncertainty by examining the implications of the system, both in the open loop and in the closed loop, to perturbation in the system matrices. This is considered by investigating the condition number of the appropriate matrices. For example, it is possible to obtain a matrix transfer function between the desired and actual outputs. If this is poorly conditioned then the effects of parameter perturbation will be much greater than if it is well conditioned. This technique generates a set of metrics that give an indication of the robustness of the control strategy to parameter perturbation.

This is a computational approach from numerical analysis, rather than the statistical approach frequently used to consider uncertainty. The condition number of a linear transfer system is based upon vector norms, and so this approach is similar to that used in robust control. In optimal $\mathcal{H}_\infty$ control, for example, the objective is to minimise the parameter uncertainty that can destabilise a system using the ∞-norm as the metric. The ∞-norm is often a measure of the 'worst case scenario', and thus the controller and the uncertain process can be considered to be players in a conventional game theoretic analysis. This is an example of a technique widely used in the study of economic problems, and used in economic control type experiments, being incorporated into the general control community.

The work widely uses vector norms as metrics. The main linear optimal control used is linear quadratic optimal control. In this case, the objective function is a weighted 2-norm. Further, as with most feedback control, the control rule is heavily dependent upon the system parameters, but, unlike adaptive or $\mathcal{H}_\infty$ control, parameter uncertainty is not explicitly incorporated into the rule generation. For the cost of explicitly including parameter uncertainty is more complex rules. The implications of parameter uncertainty can be considered by examining metrics that measure the effects of this uncertainty. As the condition number of a matrix is a multiplier metric of the effects of perturbations on the matrix, then those matrices (including matrix transfer functions) that are well-conditioned are more robust to parameter uncertainty than to those that are ill-conditioned. Different condition numbers for the same matrix results from using different vector norms; and while all vector norms are related, each measures a slightly different characteristic, so it is worthwhile to consider a range of such norms.

The approach used to uncertainty in this work differs from the usual economic approach of considering the statistical properties of the uncertainty. It is also more general than that frequently used in the robust control literature where the type and measure of uncertainty are specified. It is, though, more difficult to be rigorously incorporated into the design of a control rule.

Chapter 5 is concerned with software for the simulations and control of linear state space models. The Chapter introduces the use of a visual simulation environment and shows how these tools can be used to implement the ideas of earlier Chapters. This type of visual simulation, while common in the control engineering community, has received little attention by economists. The Chapter presents a set of software tools whereby policy analysis through control experiments using the techniques of this work can be easily undertaken. The software environment also allows for the visual description of the techniques.

With Chapter 6 the focus turns to nonlinear models; and in that Chapter a review is undertaken of a number of recent applications of control theory to macroeconomic policy analysis. The applications have followed two main approaches; with, firstly, the approach of linearisation of the nonlinear model and the use of linear quadratic optimal control techniques; and, secondly, the approach of nonlinear optimisation. Both these approaches differ from the approach of this work where separate model based control is used. The only similar approach to that used in this work is a study of nearly two decades ago by Zarrop, Holly, Rüstem, Westcott and O'Connell (1979). That study is also reviewed in the Chapter.

Chapter 7 constructs a large nonlinear macroeconometric model for use in control experiments. An aim of this work is to use best practise econometric models developed in the usual manner by econometricians. For this reason, the Murphy model, which is a major Australian dynamic macroeconometric model, is used. In the Chapter, the published equations and coefficients of the model are used 'as is', and the model is solved. The model is, though, transformed into a nonlinear state space form; and is also incorporated into the visual simulation environment. Again, this approach is novel to the solution of large macroeconometric models.

Chapter 8 undertakes policy analysis experiments on the high order nonlinear macroeconometric model using the low order linear model based controllers. Here the low order model learns to behave like the high order system for which it is generating control policies through output-injection learning. This is a low order linear observer for a high order nonlinear model. The control techniques developed in earlier Chapters are used in these experiments. Many of the other techniques of earlier Chapters are also used, including the robustness to parameter uncertainty using condition number metrics, and visual simulation. The Chapter presents new ways of applying control theory to economics.

While Chapter 8 used linear observers, Chapter 9 is concerned with nonlinear observers. The Chapter extends the conventional linear observer into a nonlinear observer. In the Chapter, a nonlinear model is used with the addition of an output-injection learning term, to form the observer. State feedback gain matrices and the learning error gain matrices are computed on the basis of a linearisation of the nonlinear model. This linearisation may be undertaken periodically for improved performance. The approach is illustrated with a low order highly nonlinear model. A low order, continuous time, nonlinear observer is also used to learn about, and develop control policies for, the high order, discrete time, practical macroeconometric model. This is a novel approach to control policy design with economic systems.

The final Chapter, Chapter 10, draws some conclusions from this research and suggests areas that could benefit from further investigation.

Appendix A discusses the software engineering of this research project. The body of this work is concerned with the presentation of concepts and tools, but a large software engineering effort was undertaken in the production of these results. A brief overview of the software engineering component of the project, is presented in Appendix A.

Chapter 2

OBSERVERS FOR LINEAR RATIONAL EXPECTATIONS MODELS

2.1 Introduction

The aim of this Chapter is to show that standard control techniques (including optimal control) can be used with linear models containing rational expectations. It is argued that the major problem with rational expectations is that the initial values of the expectations states are unknown. This Chapter shows how an observer can be used to learn these states and at the same time be used to calculate control policies. Further, the basic aim of rational expectations is to allow for variables that can 'jump' to 'news'. It is also shown that the observer can learn these disturbances.

Economists have long considered economic actors expectations to be important. Expectations were formally introduced into economic models in the form of either adaptive or backward looking expectations. In either case expectation variables are based upon previous variable values. The Lucas (1976) criticism argues that if economic actors are rational and thus do not make systematic errors then their expectations will change with variations in economic policy. Thus models that contain expectations based only on past behaviour (or no expectations) will not reflect the true nature of variations in economic policy, and thus should not be used in its development. Lucas (1976) argued that fixed parameter extrapolative expectations equations are inadequate, and even misleading, for policy analysis.

One response to the Lucas (1976) criticism was the incorporation of forward looking expectations into economic models. Forward looking economic actors are those that consider the future behaviour of economic variables, and react according to how they expect these economic variables to behave. This leads to variables in economic models that are free to 'jump' in response to 'news'. The rational expectations hypothesis states that forward looking economic actors are not consistently wrong in their expectations of economic variables, so that, on average, the expected value of some future value of the variable is its actual value. In particular, the strong form of the rational expectations hypothesis argues that economic actors have complete information about the model and thus know its functional form and parameters. In this case, the actors expectations will be the models own predictions, and the expectations are said to be model consistent. Under this hypothesis, the expected value of a variable becomes its mathematical expectation, allowing for expectations variables to be solved recursively if initial values are known. But the central problem is that the initial values of expectations variables are unknown.

Models with rational expectations can result in optimal policies becoming time-inconsistent. This is the Kyland and Prescott (1977) criticism of the use of linear quadratic optimal control theory with linear rational expectations models. The optimal policy is generated for a period on model consistent expectations. The expectations then change on the basis of the announced policy, and so the policies will no longer be optimal from the next period onwards. This gives rise to an incentive for policy makers to renege on the rule. Kyland and Prescott (1977) strongly argued that this refuted the use of linear quadratic optimal control theory with economic models and argued instead for the use of fixed (i.e. precommited) open loop rules, such as a fixed percentage increase in the money supply.

The harsh criticism of Kyland and Prescott (1977) has weakened with time (Begg, 1982; Pagan, 1996; Petit, 1990), but the time inconsistency problem still remains. In this Chapter, the approaches of Currie and Levine (1993*c*) and McKibbin and Sachs (1991) to overcome the problem are examined. Currie and Levine (1993*c*) attempt to build, with linear quadratic optimal control theory, fixed normative (i.e. time invariant) state feedback rules. These, they argue, have the advantage of precommitment by policy makers, and also have flexibility to respond to changing circumstances. Unfortunately time inconsistency stops Currie and Levine (1993*c*) from building such a rule. In this Chapter it is shown how such a rule can be built using an observer.

McKibbin and Sachs (1991) use linear quadratic optimal control theory in the analysis of inter-economy co-operation. They, too, are subject to the problem of unknown initial values for expectations states. They use an estimation procedure to estimate the initial state values.

2.2 The Currie & Levine Approach

Currie & Levine (Currie and Levine, 1993*c*) are concerned with the development of feedback rules for linear stochastic rational expectations models using linear quadratic optimal control techniques. They apply their approach to a number of small models, and also to a linearised version of the London Business School econometric model (Christodoulakis, Levine and Gaines, 1993). The main problem that Currie & Levine are concerned with is the development of simple time-consistent rules. Time-consistent rules are those such that re-optimisation at a later date would not lead to a different rule whereby policy makers would be encouraged to renege on their initial optimal plans and introduce a new set. They are primarily searching for normative or fixed numerical rules based upon the current state as, they argue (Currie and Levine, 1993*b*, p.98), this is time-consistent and has the advantage of enforcing precommitment upon policy makers. They are thus searching for time invariant state feedback gains. Unfortunately, though, their approach leads to the development of an optimal control rule "with memory" (Currie and Levine, 1993*a*, p.127) as it includes an integral term. Thus a different fixed rule will be obtained from whenever is taken to be the initial time period for the integrator, so that optimisation at later periods results in different optimal rules and thus gives time-inconsistent policies. They also consider "quasi-optimal rules" where the integral term is not included in the control rule, and argue that if the "world" is sufficiently stochastic, then these sub-optimal rules may be best.

The Currie & Levine approach will be considered in detail for the aim is to show that an observer based technique can obtain the desired results, is much less complicated and more mathematically robust. The Currie & Levine approach is also illustrative of the economics literature on linear quadratic optimal control applications. In the following, the nomenculture is that consistent with this book rather than that used by Currie & Levine.

Currie & Levine consider the economy to be in a stable long run equilibrium but to be subject to shocks, and use linear quadratic optimal control theory to return the economy to the long run equilibrium. Continuous time models are used to model the adjustment of the economy, and this compares to the usual economic approach of discrete time models. For their stochastic rational expectations models, they consider the economy to be modelled by a state equation of the form

$$\begin{bmatrix} d\mathbf{w}(t) \\ d\mathbf{z}^{\mathbf{e}}(t) \end{bmatrix} = A \begin{bmatrix} \mathbf{w}(t) \\ \mathbf{z}^{\mathbf{e}}(t) \end{bmatrix} dt + B\,\mathbf{u}(t)dt + d\mathbf{v}(t) \tag{2.1}$$

where $\mathbf{w}$ is a vector of known state variables with given initial values; $\mathbf{z}^{\mathbf{e}}$ is a vector of unknown state variables that can freely 'jump' to 'news'; $\mathbf{u}$ is a vector of control instruments; and $\mathbf{v}$ is a vector of white gaussian noise which shocks the system. The matrices A and B are appropriately dimensioned and are time invariant, with the system $(A,\ B)$ being assumed to be controllable. All variables are measured as deviations from the steady state solution. The superscript e refers to expectations variables, and by the rational expectations hypothesis, these forward looking variables will be equal to their mathematical expectations which is their actual values, so that $\mathbf{z}^{\mathbf{e}} = \mathbf{z}$.

Currie & Levine are then concerned with developing a set of optimal policies that aim to return this system back to its stable steady state. They basically consider the Linear Quadratic Regulator problem, with an objective (or social loss) function of the form

$$J = \frac{1}{2}\int_{t_0}^{t_f} e^{-\rho t}(\mathbf{x}(t)^T Q\mathbf{x}(t) + 2\mathbf{x}(t)^T U\mathbf{u}(t) + \mathbf{u}(t)^T R\mathbf{u}(t))\, dt \tag{2.2}$$

where T is the transpose operator; $\mathbf{x} = [\mathbf{w}^T, \mathbf{z}^{\mathbf{e}T}]^T$; Q, U and R are time invariant matrices with R being symmetric and positive definite and Q being symmetric and non-negative definite. Future values are discounted by the discount rate ρ $(\rho > 0)$ over the time horizon from t_0 to t_f. Currie & Levine aim to develop normative feedback rules which are fixed in the sense that they are a time invariant multiple of the current state values. These rules are obtained by using an infinite time horizon in the social loss function (as t_f approaches ∞).

The model (Equation 2.1) is stochastic. Currie and Levine (1993*b*) first consider the deterministic case and then show that, for the rules which they generate, the certainty equivalence principle holds. Thus the optimal control for the stochastic case is equivalent to that under the deterministic case. The certainty equivalence principle is applicable as the shocks to the model are independent white gaussian noise.

Solving the problem of minimising the social loss function subject to the economic system by Pontryagin's minimum principle, they obtain a coupled set of state and co-state equations which can be given by the Hamiltonian Equation[1]

$$\begin{bmatrix} \dot{\mathbf{x}}(t) \\ \dot{\mathbf{p}}(t) \end{bmatrix} = \begin{bmatrix} A - BR^{-1}U^T & -BR^{-1}B^T \\ UR^{-1}U^T - Q & \rho I - A^T + UR^{-1}B^T \end{bmatrix} \begin{bmatrix} \mathbf{x}(t) \\ \mathbf{p}(t) \end{bmatrix} \tag{2.3}$$

$$= H \begin{bmatrix} \mathbf{x}(t) \\ \mathbf{p}(t) \end{bmatrix} \tag{2.4}$$

where $\mathbf{p} = e^{\rho t}\lambda^T$ for the Lagrangian multipliers λ. In the usual problem, where there are no expectations states, this differential equation is a two point boundary value problem with initial conditions for the states and terminal conditions for the co-states. The equation (Equation 2.3) can be solved and the optimal control calculated. The problem, for the rational expectations models, lies with the boundary conditions. The terminal condition for the co-state variables is given ($\mathbf{p}(t_f) = \mathbf{0}$), and the initial conditions of the non-expectations states are also given ($\mathbf{w}(t_0) = \mathbf{w}_0$), but this is not enough boundary conditions to solve the equation. Currie & Levine obtain the remaining boundary conditions by partitioning the co-state variables into two: those ($\mathbf{p_1}$) associated with the predetermined non-expectations state variables ($\mathbf{w}$), and those ($\mathbf{p_2}$) associated with the expectations state variables ($\mathbf{z}$). They specify that the co-state expectations variables have an initial condition of zero (i.e. $\mathbf{p_2}(0) = \mathbf{0}$). The Hamiltonian Equation then becomes solvable, but the optimal control rule will contain both state and co-state variables.

[1] The derivation of the Hamiltonian Equation for a similar problem is given in the Appendix (Section 2.7 on page 60) of the Chapter.

Currie & Levine proceed as follows: they repartition the Hamiltonian Matrix, (i.e. H in Equation 2.4), into those variables that are predetermined, (i.e. $[\mathbf{w}^T, \mathbf{p_2}^T]^T$), and those that are not predetermined, (i.e. $[\mathbf{p_1}^T, \mathbf{z}^T]^T$), as this forms the model into the conventional form for rational expectations models. They then consider the infinite time horizon case ($t_f = \infty$), where the transversality condition holds and use the Blanchard and Kahn (1980) condition for obtaining a unique solution to a linear rational expectations model. The Blanchard and Kahn (1980) condition requires the saddlepath property of the driving matrix so that there must be as many unstable eigenvalues as there are expectations variables. For Currie & Levine, as long as the repartitioned Hamiltonian Matrix, H, has the saddlepath property then model stability imposes that the non-predetermined variables can be written in terms of the predetermined variables, as

$$\begin{bmatrix} \mathbf{p_1}(t) \\ \mathbf{z^e}(t) \end{bmatrix} = -M_{22}^{-1}M_{21}\begin{bmatrix} \mathbf{w}(t) \\ \mathbf{p_2}(t) \end{bmatrix} = -N\begin{bmatrix} \mathbf{w}(t) \\ \mathbf{p_2}(t) \end{bmatrix} = -\begin{bmatrix} N_{11} & N_{12} \\ N_{21} & N_{22} \end{bmatrix}\begin{bmatrix} \mathbf{w}(t) \\ \mathbf{p_2}(t) \end{bmatrix} \tag{2.5}$$

where M diagonalises the Hamiltonian Matrix (H) and M_{22} is the Jordan block associated with the unstable eigenvalues, and M_{21} is the Jordan block associated with the stable eigenvalues (Christodoulakis, 1989*a*). Then, provided the discount rate, ρ, is small, the optimal control is given by

$$\mathbf{u}(t) = D\begin{bmatrix} \mathbf{w}(t) \\ \mathbf{p_2}(t) \end{bmatrix} \tag{2.6}$$

where

$$D = -R^{-1}\left[\, U_1 - B_1N_{11} - U_2N_{21} \quad B_2 - B_1N_{12} - U_2N_{22} \,\right] = \left[\, D_1 \quad D_2 \,\right] \tag{2.7}$$

and all matrices are conformably partitioned with the Hamiltonian Matrix.

This gives a fixed optimal control rule in terms of the current non-expectations states and the current expectations co-states. To remove the expectations co-states, Currie & Levine solve the differential equation for the expectations co-states (from Equation 2.3) to obtain their "simple 'quasi-optimal' rule with precommitment" (Currie and Levine, 1993*b*, p.103) of

$$\mathbf{u}(t) = D_1\mathbf{w}(t) + D_2\int_{t_0}^{t} e^{C_{22}(t-\tau)}C_{21}\mathbf{w}(\tau)\,d\tau \tag{2.8}$$

where $C = H_{11} - H_{12}N$, and all matrices are conformably partitioned.

The optimal control rule is now expressed in terms of the known states, but includes an integral term as well as a proportional term. The integral term leads to time inconsistency as different starting points for the integral will lead to a different value for the optimal control, even with the states ($\mathbf{w}$) having the same values in both situations. That is, re-optimising at the time t_1 rather than t_0, may cause the control rule for time t to be different because the integral is calculated over $[t_1, t_f]$ rather than $[t_0, t_f]$. Further, the rule is based upon the initial conditions of the co-state variables associated with the expectations states ($\mathbf{p_2}$) being zero. But generally, for $t > t_0$ this will not the case, so that re-optimising at time $t_1 > t_0$, sets $\mathbf{p_2}(t_1) = 0$ which means that policies formed at t_0 are reneged on. Now, by rational expectations, expectations are based upon being formed consistently with the rule, but if they are based upon reneging, then the rule is not optimal as the assumption of rational expectations is broken. The rule must be credible with the expectation formers to be optimal, which is an issue considered in Currie and Levine (1993*a*).

The above rule depends upon the saddlepath property of the Hamiltonian Matrix, and Currie & Levine also show that models where the number of stable eigenvalues is greater than or less than the number of expectations variables lead to sub-optimal results. To ensure the saddlepath property Currie & Levine argue that as the system (A, B) is controllable, then a D can be found so that the matrix $[A + BD]$ has arbitrary eigenvalues, and, in particular, ensure that the saddlepath property holds. Thus they use standard eigenvalue (or pole) placement techniques to ensure that the saddlepath property holds.

Because the feedback rule is time-inconsistent, and because of its "potential complexity", Currie and Levine (1993*b*, p.104) consider rules of the type

$$\mathbf{u}(t) = \mathcal{D}_1 \mathbf{w}(t) \tag{2.9}$$

Further, they assume that the feedback gains matrix, $\mathcal{D}_1$, which are only based on the non-expectations states, is chosen such that the closed loop state equation driving matrix $[A + B[\mathcal{D}_1, O]]$ has the saddlepath property. Here O is the square matrix of zeros of dimensionality of number of expectations states. Currie and Levine (1993*a*, p.129) argue that these expectations states have the solution

$$\mathbf{z}^{\mathbf{e}}(t) = -\mathcal{N}(\mathcal{D}_1)\mathbf{w}(t) \tag{2.10}$$

For a given $\mathcal{D}_1$, say $\mathcal{D}_1^{(0)}$,

$$\mathcal{N}(\mathcal{D}_1^{(0)}) = \mathcal{N} = \mathcal{M}_{22}^{-1}\mathcal{M}_{22} \tag{2.11}$$

where

$$\mathcal{M} = \begin{bmatrix} \mathcal{M}_{11} & \mathcal{M}_{12} \\ \mathcal{M}_{21} & \mathcal{M}_{22} \end{bmatrix} \tag{2.12}$$

and, again, $\mathcal{M}$ is conformably partioned with the non-expectations and expectations states, and is the matrix of left eigenvectors of the closed loop state equation driving matrix.

Currie and Levine (1993*a*, p.129) then substitute for the expectations states using Equation 2.10, giving a social loss function of

$$J = \frac{1}{2}\int_{t_0}^{\infty} e^{-\rho t}(\mathbf{w}(t)^T\tilde{Q}\mathbf{w}(t) + 2\mathbf{w}(t)^T\tilde{U}\mathbf{u}(t) + \mathbf{u}(t)^T R\mathbf{u}(t))\, dt \tag{2.13}$$

They then obtain the following "standard" solution for the optimal control gains of

$$\mathcal{D}_1^{(1)} = -R^{-1}(B_1^T S^{(1)} + \tilde{U}^T) \tag{2.14}$$

where $S^{(1)}$ is the solution to the Algebraic Riccati Equation

$$S^{(1)}(\tilde{A} - \frac{\rho}{2}I) + (\tilde{A} - \frac{\rho}{2}I)^T S^{(1)} + \tilde{Q} - (SB_1 + \tilde{U})R^{-1}(B_1^T S^{(1)} + \tilde{U}^T) = \mathbf{0} \tag{2.15}$$

The subscripts refer to conformably partitioned matrices according to non-expectations and expectations states, and

$$\tilde{A} = A_{11} - A_{12}\mathcal{N} \qquad \tilde{U} = U_1 - \mathcal{N}^T U_2$$

$$\tilde{Q} = Q_{11} - N^T Q_{12} - Q_{21}N + N^T Q_{22}N$$

$$Q = \begin{bmatrix} Q_{11} & Q_{12} \\ Q_{21} & Q_{22} \end{bmatrix} \qquad U = \begin{bmatrix} U_1 \\ U_2 \end{bmatrix}$$

The problem is that $\mathcal{N}$ is a function of $\mathcal{D}_1$. This is because Currie & Levine use eigenvalue placement feedback to ensure their model has the saddlepath property, so that they can then use optimal control feedback to determine the control rule. The feedback gains are then dependent upon each other. Currie and Levine (1993*a*, p.129) argue this function can be considered to be a "reaction function" for the forward looking actors "capturing" their "response" to a "policy announcement". "If the controller were to exploit its knowledge of this reaction function, then it would be acting as a *Stackelberg leader*" (Currie and Levine, 1993*a*, p.130). But the resultant rule from this game is time-inconsistent.

The optimal control gain Equation 2.14 generates a sequence for $\mathcal{N}$. Currie and Levine (1993*a*, p.131) state: "we have not been able to derive any convergence properties for this sequence analytically". For the models they use, they find convergence, "provided a suitable initial rule $D_1^{(0)}$ is chosen" (Currie and Levine, 1993*a*, p.131).

Currie and Levine (1993*b*) show that certainty equivalence holds for this rule as well as the rule with the integrator given above, but does not necessarily hold for other more simple rules.

In summary, then, Currie & Levine are searching for simple normative (time invariant) rules for use with stochastic rational expectations models. They argue that such a rule has the advantages of time consistency, precommitment, and is more flexible than a fixed rule with no feedback. The problem, though, is that the initial values of the expectations variables are unknown. To overcome this they develop a number of feedback rules. The first results in an integral term in the control rule, and is thus time-inconsistent. Their second rule is also complex, and is restricted in its usage. In Section 2.4.2 (on page 32) of this Chapter, a feedback rule is developed that achieves the aim of Currie & Levine.

2.3 The McKibbin & Sachs Approach

McKibbin & Sachs (McKibbin, 1987; McKibbin, 1989; McKibbin, 1993; McKibbin, 1996; McKibbin and Sachs, 1988; McKibbin and Sachs, 1989; McKibbin and Sachs, 1991; Sachs and McKibbin, 1985) also use optimal control theory with linear stochastic rational expectations models. As with Currie & Levine, they basically consider the Linear Quadratic

Regulator problem, but they differ in that they solve the problem in discrete time. They are also concerned with multi-economy modelling and use their models to examine such questions as the advantages of global co-operation compared with individual economy optimisation, so that the problem is often formulated as a dynamic Nash linear quadratic game.

The McKibbin & Sachs approach is used as an illustration for it shows a different application of linear quadratic control theory with linear rational expectations models; it uses the more usual form in economics of solving the problem in discrete time[2] and using dynamic programming, though it is unusual in that it uses an output equation; and, it solves the problem of the unknown initial values of the expectations variables by estimating them. On this latter point, McKibbin & Sachs use an iterative procedure for the estimating and they then solve their models in the usual recursive manner.

The McKibbin & Sachs models may[3] be written with a state equation of the form

$$\begin{bmatrix} \mathbf{w}_{t+1} \\ \mathbf{z^e}_{t+1} \end{bmatrix} = A \begin{bmatrix} \mathbf{w}_t \\ \mathbf{z^e}_t \end{bmatrix} + B\mathbf{u}_t + G\mathbf{e}_t + F\mathbf{v}_t \tag{2.16}$$

and an output equation of the form

$$\mathbf{y}_t = M \begin{bmatrix} \mathbf{w}_t \\ \mathbf{z^e}_t \end{bmatrix} + D\mathbf{u}_t + E\mathbf{e}_t + L\mathbf{v}_t \tag{2.17}$$

where the subscript t refers to discrete time periods; $\mathbf{w}$ is a vector of non-expectations state variables; $\mathbf{z^e}$ is a vector of expectation state variables, referred to as "jumping" variables by McKibbin & Sachs; $\mathbf{u}$ is a vector of control variables; $\mathbf{e}$ is a vector of other exogenous variables; $\mathbf{v}$ is a vector of white gaussian noise; and $\mathbf{y}$ is a vector of output variables. The matrices A, B, D, E, F, G, L and M are appropriately dimensioned and time invariant.

[2] The McKibbin-Sachs Global (MSG) model (McKibbin and Sachs, 1991) is presented as a nonlinear continuous time model which is linearised and converted into a discrete form for the analysis.

[3] Again the nomenculture is that used in this book rather than that used by McKibbin & Sachs. For clarity in this book, many of the major components have been given the same names. So, for example, the state equation driving matrix has been called A in both Equations 2.1 and 2.16. This is not meant to infer that it is the same matrix.

The initial values for $\mathbf{w}$ are known and $\mathbf{e}_t$ is known for all relevant time periods. By the rational expectations hypothesis, the expected value of the expectations states is their mathematical expectation, so that $\mathbf{z}^\mathbf{e} = \mathbf{z}$. But, again, the initial values are not known, so the model cannot be solved recursively.

McKibbin & Sachs use a quadratic objective function; and, as the noise is white and gaussian in the model (Equations 2.16 and 2.17), the certainty equivalence principle is used so that the deterministic solution is equivalent to the stochastic solution. The objective function is

$$J = \sum_{t=0}^{\infty} \rho^t (\mathbf{y}_t^T Q \mathbf{y}_t) \tag{2.18}$$

where ρ is the discount rate, and Q the (time invariant) weighting matrix. Note that this objective function places no penalties on the controls, and, as a regulator, will be minimised when the outputs are zero. Unlike Currie & Levine, who express their models in deviations, McKibbin & Sachs express their models in levels.

Again, the major problem is with the unknown expectations states. Basically, the McKibbin & Sachs approach is to develop a method so that the initial values of these states can be estimated and use the rational expectations hypothesis to solve the model. Firstly, they assume terminal values for the expectations variables, for, after some distant time period ($t = N$), it is assumed that expectations are irrelevant. Then, secondly, they assume that at any time period the expectations states are a linear function of the non-expectations states, of other predetermined (exogenous) variables, and the future time path of these predetermined variables. They write this function as

$$\mathbf{z}_t = H\mathbf{w}_t + K\mathbf{e}_t + C_t \sum_{i=1}^{N} \mathbf{e}_{t+i} \tag{2.19}$$

Terminal values for period N are assumed, so that Equation 2.19 can be run out backwards in time. The same saddlepath (or stable manifold) property for the model is assumed as with Currie & Levine, so that the model is assumed to have as many unstable eigenvalues as expectations variables. Thus this equation for expectations states is stable along this trajectory. The equation is iterated (with the control generated and the model solved for each time period) until it converges for a fixed point for H and K. These matrices along with the time varying set of matrices C_t can be used to give the values of the expectations states for any time period by using Equation 2.19. To consider the implications of the length of the time horizon on these matrices, McKibbin & Sachs vary N until H, K and C_t still converge to a unique fixed point. Thus there is a fixed point iterative procedure both to determine these matrices and the terminal period N. This multiple fixed point iterative procedure is of the same type as the frequently used Fair and Taylor (1983) solution for rational expectations models.

The model can now be run out forwards in time for the non-expectations states have known initial values, and the expectations states values can be calculated from Equation 2.19. For the calculation of the optimal control McKibbin & Sachs use dynamic programming. The technique for the calculation of the initial states has similarities to the dynamic programming approach.

2.4 An Observer To Learn Expectations

The problem with forward looking economic actors is that it leads to models that contain expectations variables that are free to 'jump' to 'news'. The rational expectations hypothesis limits these variables so that the models can be solved recursively, but there still remains the problem that initial values of these states are unknown. This is particularly a problem with linear quadratic optimal control and was used by Kyland and Prescott (1977) to refute its use as an economic policy development tool.

The Currie & Levine and McKibbin & Sachs approaches considered above, illustrate techniques that have been developed to overcome these problems. Both approaches have their limitations and require simplifying assumptions. Currie & Levine aimed for fixed normative feedback rules and, when limited to the non-expectations states, found that these could not be generated because the initial values of the expectations states were unknown. They have an analysis that leads to a complicated algorithm for the control rule. McKibbin & Sachs simply estimate the unknown initial values of the expectations states, and then, effectively, expectations are part of their model. Their approach really does not leave the expectations states free to 'jump'. Both approaches also rely on numerical techniques with no analysis of the implications of numerical issues.

In this Section an observer based approach is proposed for learning the expectations states, and then these learned states are used in the development of linear quadratic optimal control. This is a simpler approach and has the advantage that it is a standard control technique (Luenberger, 1971; Åström and Wittenmark, 1990; Ogata, 1990). The aim of the observer is to learn the expectations states, both their initial values and when they 'jump' to 'news'.

With the rational expectations hypothesis, it is assumed that economic actors have full and complete knowledge of the model, so that expectations can be considered as being generated by that model. With a weaker form of the hypothesis, the expectations can also be considered as disturbances as they 'jump' to 'news' from outside the model. Here an observer is built to learn the expectations. The observer can have complete knowledge of the model by the rational expectations hypothesis. The observer learns the expectations generated by the model.

Another view is to consider that the model is representing the 'real economy', and, in the 'real economy', the expectations states do have actual values. The problem is that they are unknown to the model. Given the rational expectations hypothesis, once the initial values for the expectations variables are known the model can be solved. The approach here is to try and learn the states for the expectations variables from values that are occurring in the 'real economy'. It is assumed that the model is a good representation of the 'real economy', and the aim is to build an observer that can learn the expectations states for the purposes of generating the optimal controls. The 'real economy' is represented by the model but it is also subject to shocks and disturbances. Thus the expectations can be generated by the model (plus shocks and disturbances) and it is the objective of the observer to learn them for the purposes of developing control policies.

McKibbin & Sachs also aim to recreate the initial values of the expectations states by making some terminal value assumptions and iterating backwards. This compares to the approach here whereby, rather than trying to estimate the initial values of the expectations states, an observer is built that learns to behave like the model. The problem then becomes whether or not the observer model can then be used to generate control policies, and it is shown below that the problem of determining the control is also independent from the problem of observing the states.

One significant advantage of building an observer is that it can be extended for control methodologies without altering the economic structure of the original model. This is illustrated in Chapter 4.

Another significant advantage of the observer based approach is that it does not rely on the saddlepath property of the model, which requires that there must be exactly the same number of unstable eigenvalues as there are expectations variables in the state equation driving matrix (A in both Equations 2.1 and 2.16). In the approach presented here, it does not matter whether the model is stable, has the saddlepath property, or is completely unstable. The technique will develop control policies in any case, provided the model is stabilisable. This is illustrated in Section 2.5.3 (on page 45).

The basic approach of both McKibbin & Sachs and Currie & Levine is that of a Two Point Boundary Value Problem with known initial values of the non-expectations states and known terminal values of the expectations states. To be able to solve their problem, it is necessary to have a saddlepath and to be able to stay upon the stable arm (or manifold) of the saddlepath. In the observer based approach, the problem is considered as an initial value problem with unknown values of the expectations states. The problem is to learn the expectations states. Further, expectations states can be considered as being disturbed as they 'jump' to exogenous 'news'. Expectations states are not significantly different from non-expectations states, and feedback rules can be based upon all states.

2.4.1 Methodology

The approach here will be considered in continuous time following Currie & Levine, though the discrete case is analogous. Further, for convenience the model will have fixed parameter matrices, but the arguments still hold for time varying parameter systems. The economic model is specified with a state equation of

$$\begin{bmatrix} d\mathbf{w}(t) \\ d\mathbf{z}^{\mathbf{e}}(t) \end{bmatrix} = A \begin{bmatrix} \mathbf{w}(t) \\ \mathbf{z}^{\mathbf{e}}(t) \end{bmatrix} dt + B\,\mathbf{u}(t)\,dt + H\,d\mathbf{v_1}(t), \tag{2.20}$$

and an output equation of

$$\mathbf{y}_t = C \begin{bmatrix} \mathbf{w}(t) \\ \mathbf{z}^{\mathbf{e}}(t) \end{bmatrix} + D\mathbf{u}(t) + \mathbf{v_2}(t). \tag{2.21}$$

Here $\mathbf{w} \in \Re^{n_n}$ is a vector of non-expectations states, and $\mathbf{z}^{\mathbf{e}} \in \Re^{n_e}$ is a vector of expectations states. Call all the states $\mathbf{x} \in \Re^n$ where $\mathbf{x} = [\mathbf{w}^T, \mathbf{z}^{\mathbf{e}T}]^T$, and $n = n_n + n_e$. Initial values of the non-expectations states are known, but are not known for the expectations states. In rational expectations models, the vector $\mathbf{z}^{\mathbf{e}}$ is the rational expectation of $\mathbf{z}$ so that $\mathbf{z}^{\mathbf{e}} = \mathbf{z}$. In this approach, this may be relaxed and the aim is to learn $\mathbf{z}^{\mathbf{e}}$ as it 'jumps' or is disturbed by 'news'.

The vector $\mathbf{y} \in \Re^p$ contains the outputs (or policy targets) with which policy makers are concerned. The $\mathbf{u} \in \Re^m$ is a vector of controls (or policy instruments); and $\mathbf{v_1} \in \Re^r$ and $\mathbf{v_2} \in \Re^p$ are uncorrelated vectors of white gaussian noise that shock the system. Vector $\mathbf{v_1}$ represents state noise, and $\mathbf{v_2}$ is measurement noise.

The system parameter matrices are: the state equation driving matrix, $A \in \Re^{n \times n}$; the control forcing matrix, $B \in \Re^{n \times m}$; the output generation matrix, $C \in \Re^{p \times n}$; the contemporaneous control matrix, $D \in \Re^{p \times m}$; and, the state noise matrix, $H \in \Re^{n \times r}$. It is assumed that the system (A, B) is stabilisable so that the unstable eigenvalues of the state equation driving matrix are controllable (or equivalently, the uncontrollable eigenvalues are stable). It is also assumed that the system (A, C) is detectable so that the unstable eigenvalues of the state equation driving matrix are observable (or equivalently, the unobservable eigenvalues are stable) (Shahian and Hassul, 1993).

Consider, now, the problem of building an observer to learn the model's expectations states from an arbitrary initial value and after they are disturbed by 'news'. With the approach presented here it is possible to learn all the states. For ease of notation, and as later in this book, the observer technique will be used for all states. The learning of the expectations states and the learning of all states will be illustrated in Section 2.5 (on page 36).

The model (Equations 2.20 and 2.21) represents the 'real economy' or the system from which expectations are formed. The values of the expectations states (or all states in the case here) are unknown, but the outputs from this system are known. It is from these outputs ($\mathbf{y}$) that the observer learns the states.

The observer can then be formed to incorporate the difference between its output and the output from the system or 'real economy'. Using this output-injection, the observer can then be formulated as

$$\frac{d\hat{\mathbf{x}}(t)}{dt} = A\hat{\mathbf{x}}(t) + B\mathbf{u}(t) - L(\hat{\mathbf{y}}(t) - \mathbf{y}(t)) \tag{2.22}$$

$$\hat{\mathbf{y}}(t) = C\hat{\mathbf{x}}(t) + D\mathbf{u}(t) \tag{2.23}$$

where the circumflex represents observer variables (with $\hat{\mathbf{x}} \in \Re^n$ being the observer states and $\hat{\mathbf{y}} \in \Re^p$ being the observer output); and $L \in \Re^{n \times p}$ is gain matrix that allows for the altering of the emphasis of the learning component.

The observer is deterministic, as, for the analysis here, the certainty equivalence principle holds. The following, thus, only considers the deterministic case.

The aim is for the observer to learn the system's state values. Its ability to do so can be seen by considering the difference between the system's states and the observer's states. This learning error, $\tilde{\mathbf{x}} \in \Re^n$, is then given by $\tilde{\mathbf{x}}(t) = \mathbf{x}(t) - \hat{\mathbf{x}}(t)$; and the growth in the learning error is given by

$$\frac{d\tilde{\mathbf{x}}(t)}{dt} = \frac{d\mathbf{x}(t)}{dt} - \frac{d\hat{\mathbf{x}}(t)}{dt} \tag{2.24}$$

$$= A\mathbf{x}(t) + B\mathbf{u}(t) - [A\hat{\mathbf{x}}(t) + B\mathbf{u}(t) - L(\hat{\mathbf{y}}(t) - \mathbf{y}(t))] \tag{2.25}$$

$$= A\tilde{\mathbf{x}}(t) + L[C\hat{\mathbf{x}}(t) + D\mathbf{u}(t) - (C\mathbf{x}(t) + D\mathbf{u}(t))] \tag{2.26}$$

$$= [A - LC]\tilde{\mathbf{x}}(t) \tag{2.27}$$

The aim is to drive the learning error to zero, for then the observer will have completely learned the system's state values. It can be seen, from Equation 2.27, then, that the growth in the learning error is driven by the eigenvalues of the matrix $[A - LC]$, and while A and C are fixed, the gain matrix L may be chosen so that these eigenvalues have values that drives the learning error to zero as fast as desired.

The observer can be used to overcome the problem of the unknown initial values of the states as it can learn the states. For the case of expectations variables, then only some of the initial states are unknown and only they need to be learned. This is a reduced-order observer (Åström and Wittenmark, 1990). Further, in the case where the expectations states are truly free to 'jump' to 'news', then the observer can learn this disturbance. Likewise for the situation where the system is subject to stochastic shocks. Illustrations are given in Section 2.5.1 (on page 41).

There is also the issue as to whether the observer can be used in conjunction with the control of the model. Does the design of the observer effect the design of the control rule? The control rule is based on the observer but it now differs from the system in that it contains the learning term. Without loss of generality, the simple observer state feedback control rule will be used for this analysis. This control rule is of the form

$$\mathbf{u}(t) = -K\hat{\mathbf{x}}(t) \tag{2.28}$$

with $K \in \Re^{m\times n}$ is the state feedback gain. Then the closed loop form of the model will be

$$\begin{aligned} \frac{d\mathbf{x}(t)}{dt} &= A\mathbf{x}(t) - BK\hat{\mathbf{x}}(t) & (2.29)\\ &= A\mathbf{x}(t) - BK[\mathbf{x}(t) - \tilde{\mathbf{x}}(t)] & (2.30)\\ &= [A - BK]\mathbf{x}(t) + BK\tilde{\mathbf{x}}(t) & (2.31) \end{aligned}$$

Combining this with the learning error equation gives

$$\begin{bmatrix} \frac{d\mathbf{x}(t)}{dt} \\ \frac{d\tilde{\mathbf{x}}(t)}{dt} \end{bmatrix} = \begin{bmatrix} A - BK & BK \\ 0 & A - LC \end{bmatrix} \begin{bmatrix} \mathbf{x}(t) \\ \tilde{\mathbf{x}}(t) \end{bmatrix} \tag{2.32}$$

Equation 2.32 gives the combined effect of the control rule on the model and the learning of the system's states. Now, the driving matrix in the equation is block triangular, and the eigenvalues of a block triangular matrix are the union of the eigenvalues of the blocks. Thus the eigenvalues of the closed loop observer are separate from the eigenvalues associated with the observer learning the states. Hence the problem of determining the control is separate from the problem of learning the states. The design of the observer is separate from the design of the controller in the situation considered here.

In summary then, the basic problem is that the expectations states are unknown, so an observer is built to learn these states. The degree to which the observer learns about the expectations states can be set so as to drive the learning error to zero at a reasonable rate. The observer can learn the effects of, not only the initial values of the expectations states, but all states; and can learn about the stochastic shocks and disturbances such as when the expectations actually 'jump' to 'news'. The problem of designing control rules for the model is separate from that of designing the observer. The only requirements are that the expectations states (or all states in the case here) be detectable, and that the system that generates the states be fully known. This later requirement is upheld by the rational expectations hypothesis.

2.4.2 Linear Quadratic Tracking

In control applications to economics, it is usual to use optimal control rules; and, normally, linear quadratic control. McKibbin and Sachs (1991) specify the objective function (Equation 2.18) in terms of a regulator, but are unconcerned with the behaviour of the policy instruments or controls. Currie and Levine (1993*c*) are concerned with returning the model to its long run equilibrium and also use an objective function (Equation 2.2) specifying a regulator. In this book the emphasis is on policy analysis rather than policy development. The concern is whether particular policies are achievable, and what are the implications of attempting to achieve these policies. Hence the emphasis is on output trajectories specified by policy makers[4] rather than regulating about an equilibrium level. Further, policy mak-

[4] An example is the announced policy of the Reserve Bank of Australia maintaining inflation in the vicinity of 3% p.a..

ers are allowed to specify any (not just discounted) future weights for deviations between actual and desired trajectories. Finite time horizons (e.g. a political cycle) are used, and a component is included so that policy makers can emphasise, by an appropriate weight, the attainment of a final position. This produces a flexible objective function for policy analysis. By appropriate weights and desired trajectories, this objective function can also produce similar functions to those used by Currie and Levine (1993*c*) and McKibbin and Sachs (1991).

The objective function is

$$J = \frac{1}{2}\bar{\mathbf{y}}(t_f)^T Q(t_f)\bar{\mathbf{y}}(t_f) + \frac{1}{2}\int_{t_0}^{t_f} (\bar{\mathbf{y}}(t)^T Q(t)\bar{\mathbf{y}}(t) + \bar{\mathbf{u}}(t)^T R(t)\bar{\mathbf{u}}(t))\, dt \tag{2.33}$$

where the output weighting matrices $Q(t) \in \Re^{p\times p}$ can be factored into $Q(t) = Q_c(t)^T Q_c(t)$ where $Q_c(t)$ is any matrix such that (Q_c, A) is detectable; and, the control weighting matrices $R(t) \in \Re^{m\times m}$ are symmetric positive definite. The time horizon is from t_0 to t_f. The bars indicate deviations from desired policies, so that $\bar{\mathbf{y}}(t) = \mathbf{y}(t) - \mathbf{y_d}(t)$ and $\bar{\mathbf{u}}(t) = \mathbf{u}(t) - \mathbf{u_d}(t)$; where $\mathbf{y_d}(t)$ and $\mathbf{u_d}(t), t \in [t_0, t_f]$ are the (known) output trajectory desired by policy makers and the (known) desired control trajectory respectively.

This objective function specifies the Linear Quadratic Tracking Problem, rather than the Linear Quadratic Regulator Problem. The Linear Quadratic Tracking Problem was the one used in the classic Pindyck (1973) study, though there the linear model was specified as a descriptor model. The Linear Quadratic Tracking Problem is also considered by Kendrick (1981). In both these cases state tracking was used and no output equation was specified; though both studies were only concerned with a subset of the states and used zero weights on some states.

This objective function allows for the consideration of trade-offs amongst the various individual outputs over time; and, similarly, amongst the individual controls. It also allows for the investigation of the trade-offs between the controls and the outputs. If the emphasis is only upon the desired outputs, then model inversion can be used to generate a $\mathbf{u_d}$, such that the desired outputs are achieved.

There remains the issue of the generation of the optimal control rule. The optimal control generated by the Linear Quadratic Tracking Problem can be used with the observer states. This problem, for which a solution is given in the Appendix to this Chapter (Section 2.7 on page 60), creates an optimal control rule of

$$\mathbf{u}(t) = -R^{-1}(t)B^T[S(t)\hat{\mathbf{x}}(t) + \mathbf{g}(t)] + \mathbf{u_d}(t) \tag{2.34}$$

where $S(t)$ is given by the Algebraic Riccati Equation

$$\dot{S}(t) = -S(t)A - A^T S(t)R^{-1}(t)B^T S(t) - C^T Q(t)C \tag{2.35}$$

and $\mathbf{g}(t)$ is given by the predictor equation

$$\dot{\mathbf{g}}(t) = [S(t)BR^{-1}(t)B^T - A^T]\mathbf{g}(t) + C^T Q(t)\mathbf{y_d}(t) - S(t)B\mathbf{u_d}(t) \tag{2.36}$$

with the boundary conditions of

$$S(t_f) = C^T Q(t_f)C \tag{2.37}$$

$$\mathbf{g}(t_f) = -C^T Q(t_f)\mathbf{y_d}(t_f) \tag{2.38}$$

The observer based approach with this optimal control rule is illustrated in Section 2.5.2 (on page 43).

The optimal control rule given in Equation 2.34 is a time varying rule. If the weighting matrices, desired output and desired control are time invariant, then a time consistent rule of the Currie and Levine (1993*b*) type can be achieved by using the steady state values of the Riccati and predictor equations (Equations 2.35 and 2.36). This is achieved by extending the time horizon (t_f approaches ∞).

More specifically, the approach can also be used to generate the fixed time invariant rule of the type desired by Currie and Levine (1993*b*) with a social loss function of

$$J = \frac{1}{2}\int_0^{\infty} e^{-\rho t} \begin{bmatrix} \mathbf{x}(t) \\ \mathbf{u}(t) \end{bmatrix}^T \begin{bmatrix} Q & N \\ N^T & R \end{bmatrix} \begin{bmatrix} \mathbf{x}(t) \\ \mathbf{u}(t) \end{bmatrix} \tag{2.39}$$

where Q, N and R are time invariant matrices. The optimal control rule remains the same as in Equation 2.28 but the feedback gain, K, becomes

$$K = R^{-1}[B^T S + N^T] \tag{2.40}$$

and S is the positive definite solution to the modified Algebraic Riccati Equation

$$S(A - \frac{\rho}{2}I) + (A - \frac{\rho}{2}I)^T S + Q - (SB + N)R^{-1}(SB + N)^T = \mathbf{0} \tag{2.41}$$

This rule is simpler than that developed by Currie and Levine (1993c) and has the property of time consistency that they desired. The difference is that this rule is used with the observer which learns the expectations states. This rule is illustrated in Section 2.5.5 (on page 52).

2.4.3 Filtering Noise

The learning gain matrix, L, is free and can be chosen so as to drive the state learning error to zero as fast as is required. A particular case is when the aim is to specifically learn about the stochastic nature of the system, and specific knowledge is known about the stochastic shocks. In this case, a dual problem to the development of the optimal control can be set up to produce the optimal learning component. In the system (Equations 2.20 and 2.21) it was assumed that the stochastic processes that produced the state shocks and measurement errors are white, gaussian, zero mean, stationary processes with known covariances of

$$\mathcal{E}(\mathbf{v_1}) = \mathbf{0} \tag{2.42}$$

$$\mathcal{E}(\mathbf{v_2}) = \mathbf{0} \tag{2.43}$$

$$\mathcal{E}\left(\begin{bmatrix} \mathbf{v_1}(t) \\ \mathbf{v_2}(t) \end{bmatrix} [\, \mathbf{v_1}(t)^T \quad \mathbf{v_2}(t)^T \,]\right) = \begin{bmatrix} \bar{Q} & \bar{N} \\ \bar{N}^T & \bar{R} \end{bmatrix} \delta(t-\tau) \quad \forall t, \forall \tau \tag{2.44}$$

where $\mathcal{E}(.)$ is the mathematical expectations operator, $\bar{R}$ is positive definite and there exists a Q_0 such that $Q_0 Q_0^T = \bar{Q}$ and (A, Q_0) is stabilisable. Let the optimal learning gain be defined to minimise the variance of the learning error, so that

$$\bar{J} = \mathcal{E}(\tilde{\mathbf{x}}(t)^T \tilde{\mathbf{x}}(t)). \tag{2.45}$$

Then the learning gain will be given by

$$\bar{L} = (\bar{S}C^T + \bar{N})\bar{R}^{-1} \tag{2.46}$$

where $\bar{S}$ is the solution to the Algebraic Riccati Equation

$$A\bar{S} + \bar{S}A^T + H\bar{Q}H^T - (\bar{S}C^T + \bar{N})\bar{R}^{-1}(\bar{S}C^T + \bar{N})^T = \mathbf{0}. \tag{2.47}$$

The performance index, $\bar{J}$ is given by the trace of $\bar{S}$. In this case the observer is referred to as the Kalman Filter for it produces an optimal filter for the state shock noise and the measurement noise giving an optimal estimator. While the Kalman Filter can produce an optimal learning gain, it requires additional restrictions upon the system. These make it susceptible to parameter uncertainties (Doyle, Francis and Tannenbaum, 1992).

This is the dual of the time invariant Linear Quadratic Regulator Problem (i.e. using the objective function of Equation 2.39 with $\rho = 0$) with $A \Longrightarrow A^T$, $B \Longrightarrow C^T$, $Q \Longrightarrow \bar{Q}$, $R \Longrightarrow \bar{R}$, $N \Longrightarrow \bar{N}$, $K \Longrightarrow L^T$, $S \Longrightarrow \bar{S}$. When combined with optimal control is called the Linear Quadratic Gaussian Problem. It is illustrated in Section 2.5.6 (on page 54).

2.5 An Application

Many of the above points are illustrated here with the model that is used by Currie and Levine (1993*a*) when considering policy makers' credibility and the development of optimal control rules. The model is a log-linear continuous time model, and is specified in structural form as:-

$$dy = \mu[\alpha_1(e-p) - \alpha_2(r - \dot{p}^e) + \alpha_3 v - \alpha_4 s - y]\,dt + dv_1 \tag{2.48}$$

$$dv = [-\phi_1 s - \phi_2 y + \phi_3(e-p)]\,dt - dp \tag{2.49}$$

$$dp = [\beta_1 y + \beta_2(e-p)]\,dt + dv_2 \tag{2.50}$$

$$de^e = r\,dt \tag{2.51}$$

All variables are measured in terms of deviations from a given long run trend, and all coefficients (which are indicated by Greek letters) are positive. The variables have the following meanings: y is the logarithm of real output for the economy (say, real Gross Domestic Product); e is the logarithm of nominal exchange rate (measured as the price of foreign exchange); p is the logarithm of domestic prices; r is nominal interest rate (measured as a proportion); v is the logarithm of real net financial wealth of the private sector; s is autonomous taxes; and dv_1 and dv_2 are demand and supply shocks respectively. The superscript e refers to the expectation of forward looking economic actors, so that e^e is the expected exchange rate and $\dot{p}^e$ is the expected inflation rate[5] (as the price level is measured in logarithms). The first equation is an IS curve, with the economy's output adjusting sluggishly in response to competitiveness, the real expected interest rate, real wealth and autonomous taxes. The second equation is derived from the logarithmic linearisation of an equation for net financial wealth accumulation depending upon autonomous taxes, output, and competitiveness, with the rate of inflation being deducted to obtain real accumulation. The third equation is a Phillips curve, with inflation being dependent upon output and real competitiveness. The final equation, for expected exchange rate adjustment, imposes perfect capital mobility, and assumes domestic and foreign bonds are perfect substitutes.

With the strong form of the rational expectations hypothesis, economic actors are correct on average with their expectations, so that expected variables are their mathematical expectation. Hence the expected exchange rate is the actual exchange rate, and the expected inflation becomes:

$$\dot{p}^e = \beta_1 y + \beta_2(e^e - p) \tag{2.52}$$

If this (Equation 2.52) is substituted into Equation 2.48 and Equation 2.50 into Equa-

[5] The mixed notation for the first derivative of prices with respect to time follows that of Currie and Levine (1993*a*).

tion 2.49, then the model can then be written with a state equation[6] of

$$\begin{bmatrix} dy \\ dv \\ dp \\ de^e \end{bmatrix} = \begin{bmatrix} -\mu(1-\beta_1\alpha_2) & \mu\alpha_3 & \mu(\alpha_2\beta_2-\alpha_1) & \mu(\alpha_1-\alpha_2\beta_2) \\ -(\phi_2+\beta_1) & 0 & (\beta_2-\phi_3) & (\phi_3-\beta_2) \\ \beta_1 & 0 & -\beta_2 & \beta_2 \\ 0 & 0 & 0 & 0 \end{bmatrix} \begin{bmatrix} y \\ v \\ p \\ e^e \end{bmatrix} dt$$
$$+ \begin{bmatrix} -\mu\alpha_2 & -\mu\alpha_4 \\ 0 & -\phi_1 \\ 0 & 0 \\ 1 & 0 \end{bmatrix} \begin{bmatrix} r \\ s \end{bmatrix} dt + \begin{bmatrix} 1 & 0 \\ 0 & -1 \\ 0 & 1 \\ 0 & 0 \end{bmatrix} \begin{bmatrix} dv_1 \\ dv_2 \end{bmatrix} \tag{2.53}$$

which can be written as

$$\dot{\mathbf{x}}(t) = A\mathbf{x}(t) + B\mathbf{u}(t) + H\mathbf{v}(t) \tag{2.54}$$

Currie and Levine (1993*a*) do not use an output equation. They are, though, only concerned with GDP and prices in their objective function, so these states would be better considered as the outputs of the system. Here an output equation is used with their model. In some illustrations it will be assumed that there is an error in the measurement of these outputs, so a term for measurement noise is included. This leads to the following output equation

$$\begin{bmatrix} y \\ p \end{bmatrix} = \begin{bmatrix} 1 & 0 & 0 & 0 \\ 0 & 0 & 1 & 0 \end{bmatrix} \begin{bmatrix} y \\ v \\ p \\ e^e \end{bmatrix} + \begin{bmatrix} w_1 \\ w_2 \end{bmatrix} \tag{2.55}$$

which can be written as

$$\mathbf{y}(t) = C\mathbf{x}(t) + \mathbf{w}(t) \tag{2.56}$$

The model is now in the state form of Equations 2.20 and 2.21, with D the zero matrix. This state form of the model maintains the economic structure for this model. It does have an interesting feature in that the state equation driving matrix, A, has a row of zeros, and thus

[6] This equation differs from the state form used by Currie and Levine (1993*a*, p.135), firstly, in that they use e^e on the left hand side and e on the right hand side, (presumably) because $e^e = e$ by the rational expectations hypothesis; and, secondly, because the state noise matrix, H, is used here.

will always have a smallest singular value of zero. Its condition number using the 2-norm will thus be infinite. The solving of this model, in this form in the open loop, will always be an ill-conditioned problem. This reason alone makes it an interesting and difficult problem to work with, for a small perturbation to the model may lead to very different solutions.

The row of zeros will also generate an eigenvalue of zero, so the model is a Type 1 system (Ogata, 1990).

For their implementation, Currie and Levine (1993*a*) take the time unit to be one year; the demand and supply stochastic shocks are assumed to be noise with known variances and are uncorrelated, thus $dv_1 \sim N(0, \sigma_1^2 dt)$ and $dv_2 \sim N(0, \sigma_2^2 dt)$. Where measurement noise is included, it, too, is assumed to be white, gaussian[7] and independent of the demand and supply shocks. Currie and Levine (1993*a*) calibrate the model with the following parameter values: μ=0.5, α_1=0.3, α_2=0.1, α_3=1.0, α_4=0.4, ϕ_1=1.0, ϕ_2=1.3, ϕ_3=0.1, β_1=0.15, and β_2=0.25.

This results in the eigenvalues of the state equation driving matrix, A, to be $-0.26 \pm 0.83\imath$, -0.23 and 0; giving A a rank of 3. Notice the model only loosely satisfies the saddlepath property associated with the solution of rational expectations models (Blanchard and Kahn, 1980), with one *conditionally* unstable eigenvalue, and one expectations variable. The controllability matrix for the pair (A, B) and the observability matrix for the pair (A, C) both have rank of 4 and there are 4 states, so that the system is controllable and observable; and thus satisfies the requirements of the observer based approach. The singular values of A are 1.56, 0.59, 0.26, and 0.

The behaviour of the model can be seen by considering the maintained multipliers for the model variables (Challen and Hagger, 1983). These are the time responses of the states to a unit step in each input with no other effects (i.e. zero initial conditions and no noise). As the model is in deviations, this unit step is in the deviation of interest rates from its

[7] The noise should not be, correctly, described as white and gaussian, for, in continuous time, white gaussian noise has infinite variance. But for practical implementation of the model (i.e. as a discrete time approximation) the noise has known finite variance.

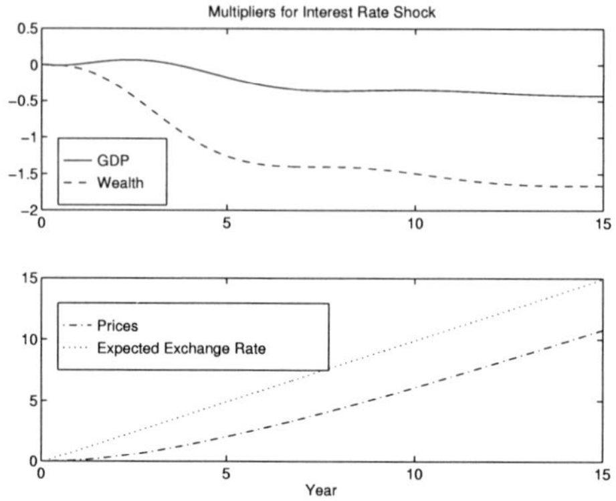

Figure 2.1: Interest Rate Shock

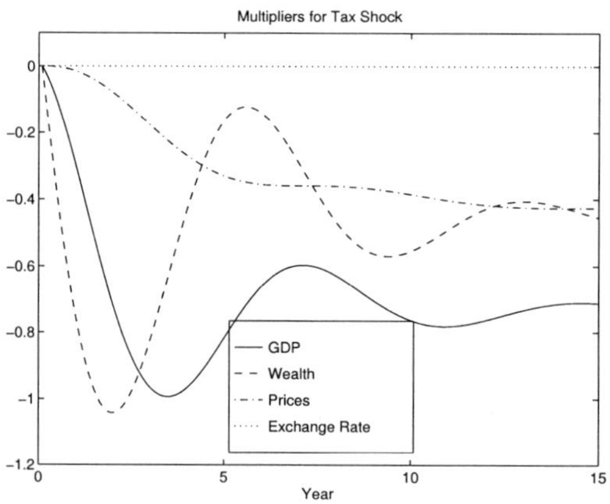

Figure 2.2: Tax Rate Shock

long-run equilibrium. The response for each control is presented in Figures 2.1 and 2.2. Figure 2.1 shows that the shock to interest rates causes the deviation in expected exchange rate to grow away from its long run equilibrium. This multiplier result is expected from Equation 2.51 (and Equation 2.53) where it can be seen that the expected exchange rate simply integrates the interest rate over time. The interest rate shock also causes prices to keep growing away from their equilibrium; whereas wealth and GDP oscillate and deviate below their long run equilibrium.

Figure 2.2 presents the multipliers for the tax shock. It confirms that an autonomous tax rate shock has no effect on the expected exchange rate; whereas it reduces wealth and GDP and thus prices. Wealth and GDP oscillate and fall away from their long run equilibrium. The dynamics of prices is similar, though with a less marked effect.

2.5.1 Learning Expectations States

For the observer based approach it is assumed that this model of Equations 2.53 and 2.55 represents the 'real' economy or it is the process by which the states are formulated. The aim is to use the observer of Equations 2.22 and 2.23 to learn the expectations states. In this illustration it is shown that the observer can learn the unknown initial values of the expectations states. This overcomes the problem for both Currie and Levine (1993*a*) and McKibbin and Sachs (1991).

Expectations states that are free to 'jump' to 'news' can also be considered as disturbances upon the system. These disturbances can also be learned. For this system, the simulation of a disturbance to the expected exchange rate can be considered by a 'jump' or shock to interest rates. The learning of this type of disturbances is also illustrated.

Finally, it is illustrated how the observer can learn the stochastic demand and supply shocks on the model. Here the system is continually effected by both demand and supply shocks rather than a single demand or supply shock as considered by Currie and Levine (1993*a*).

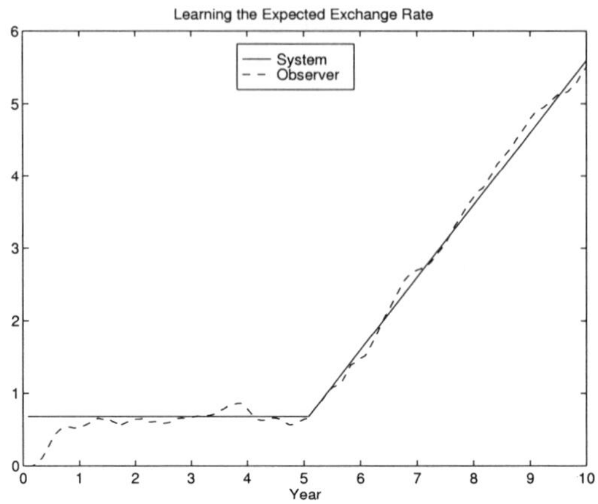

Figure 2.3: Learning the Expected Exchange Rate

The scenario for this illustration consists of the following conditions. The system is given random (between zero and one) initial values of the states; and the observer is assumed to know all the initial states except the expected exchange rate. The disturbance to the expected exchange rate is given in the form of a unit step in interest rates half way through the simulation. Following Currie and Levine (1993*a*), the demand and supply shocks are assumed to be small relative to the initial condition shock so that $\sigma_1^2 = \sigma_2^2 = 0.1$. No measurement noise is used in this illustration so that $\sigma_3^2 = \sigma_4^2 = 0$. The learning gain matrix, L, is selected so that the eigenvalues of the learning error driving matrix, $[A - LC]$, are equally spaced between -2 and -4, giving the observer a relatively slow response. The observer is allocated an initial value for the expected exchange rate state of zero, which implies that the state is at its long run equilibrium level.

Figure 2.3 illustrates the observer's ability to learn the expectations state. The Figure shows the observer tracks the expected exchange rate trajectory from the system. Firstly it learns about the difference in initial values between itself and the system for expected exchange rate, and thus is a means to overcome the problem of determining the initial value of expectations states. Secondly, it can learn about the disturbance shock to the

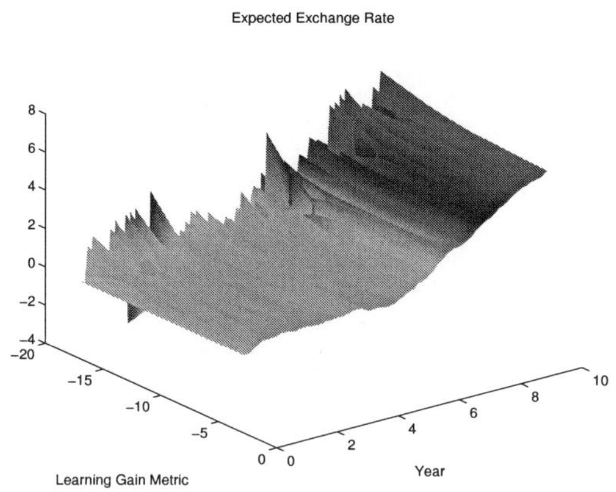

Figure 2.4: Expected Exchange Rate with Different Learning Gains

expected exchange rate. Finally, though not as obviously from Figure 2.3, it can learn about the continuous stochastic demand and supply shocks perturbing the system. The expected exchange rate has been learned through the observer's knowledge of GDP and prices. These outputs form the basis of the learning component.

By increasing the observer learning gain term, L, the observer can be made to learn the expectations states faster. Figure 2.4 shows the effects of altering the gain term. The gain metric is the average eigenvalue of the learning error driving matrix, $[A - LC]$. The learning gain, L, is chosen so that the average eigenvalue is reduced but the spread remains the same. Notice that as the learning gain is increased (so the average eigenvalue becomes more negative) the observer's response is faster in that it learns more quickly about the system. It does mean, though, that in trying to learn faster it tends to magnify the stochastic shocks.

2.5.2 *Optimal Control*

The next illustration concerns the generation of optimal control using the observer based approach. The same scenario is illustrated as that considered by Currie and Levine (1993*a*),

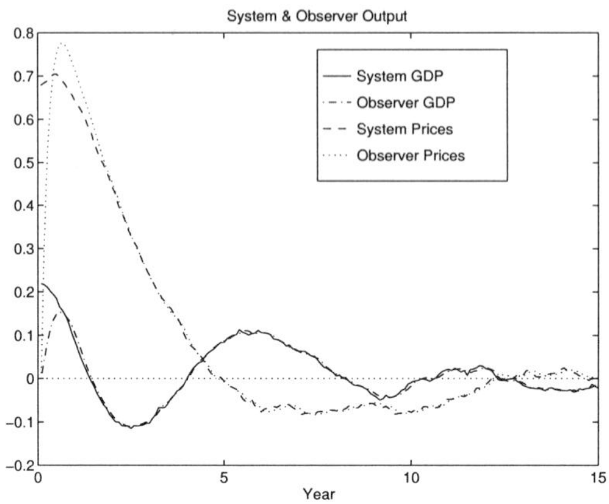

Figure 2.5: Outputs Under Optimal Control

except that demand and supply shocks occur simultaneously. The aim is to return the system to its long run equilibrium so that a desired output and control of zero is used. The objective function is that given by Equation 2.33. Time varying weights are used. Initially, the control deviations are weighted at unity, with $R(t_0)$ being the (2×2) identity matrix; and the initial output deviations being weighted at twice the controls. The weights are discounted with time with a discount rate of 5%. The demand and supply shocks (i.e. the noise) have variances of $\sigma_1^2 = \sigma_2^2 = 0.05$, but occur continually.

Again, the system is allocated random initial state values between zero and one, and, no measurement noise is used. All states are learned by the observer, and it is assumed to be initially at the long run equilibrium. The learning gain matrix, L, is, again, selected so that the eigenvalues of the learning error driving matrix, $[A - LC]$, are equally spaced between -2 and -4.

The results of this illustration are presented in Figure 2.5, where the output from both the model and the observer are plotted. Notice that both outputs (of GDP and prices) are returning towards their long run equilibrium from the initial shock, and also from the demand and supply disturbances. That is, the optimal control is stabilising the system.

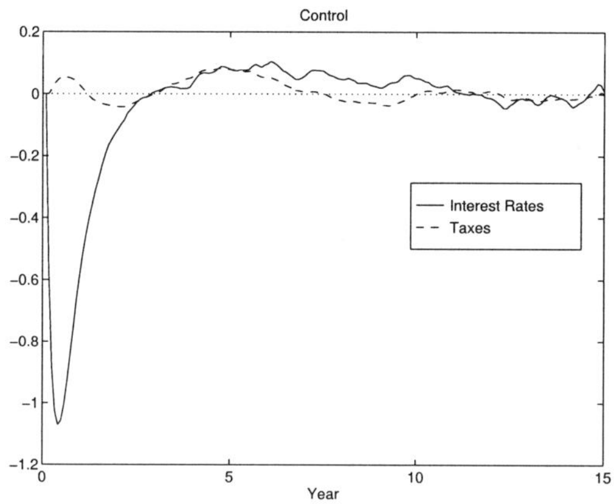

Figure 2.6: Controls Generated

Secondly, note that the observer learns the system's outputs, and comes into line with them.

This illustration shows how optimal policies can be generated under the approach presented here. Even though the initial value of the expectations state was unknown and the system is continually buffeted by stochastic demand and supply shocks, the observer can learn the system's states, including the expectations variable. At the same time it can be used to generate optimal control policies for the system. This shows that the standard control technique can be used to generate optimal policies with rational expectations models. It generates the type of rule for which Currie and Levine (1993*b*) were searching.

Figure 2.6 plots the controls generated by the optimal control rule for this illustration. Initially a cut in interest rates is generated, and then both interest rates and autonomous taxes remain close to their long run equilibrium values.

2.5.3 Saddlepath Property

Figures 2.5 and 2.6 show that the learning based approach can be used to develop optimal control rules under the scenario presented by Currie and Levine (1993*c*). For that scenario,

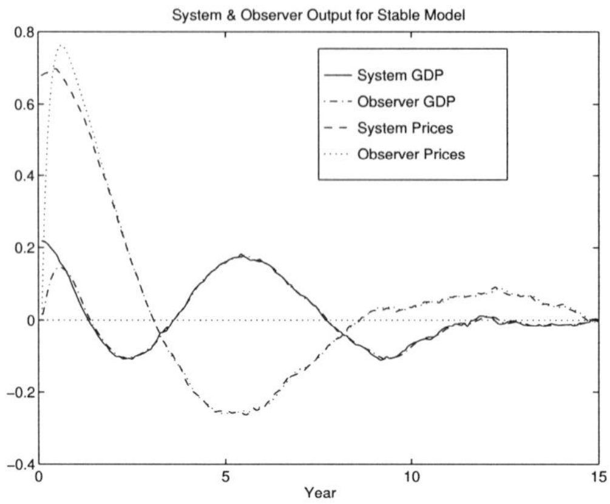

Figure 2.7: Output Using Optimal Control with Stable Model

the parameter values of the model ensured that the saddlepath property holds, for there is a single expectation state (expected exchange rate) and a single eigenvalue of the state equation driving matrix (A) with a non-negative real part. The learning based approach does not require the saddlepath property, as can be illustrated by considering a completely stable model, and by considering a highly unstable model.

With the same parameter values, the model can be made completely stable by modifying Equation 2.51 to

$$de^e = (r - p)\,dt. \tag{2.57}$$

In this case the real value of interest rates is used. This new model (Equations 2.48, 2.49, 2.50 and 2.57) results in the state equation driving matrix, A, having eigenvalues of $-0.3 \pm 0.8\imath$ and $-0.1 \pm 0.4\imath$. The saddlepath property no longer holds, and the new model is completely stable as all eigenvalues have negative real parts. Consider the generation of optimal control with the observer based approach with this modified model using the same scenario as above. Figure 2.7 presents the output under this modified model. Again, the learning based approach develops control policies that return both GDP and prices towards their long run equilibrium.

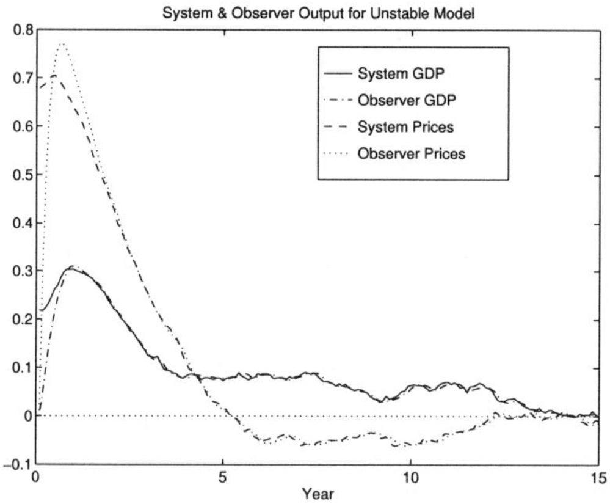

Figure 2.8: Output Using Optimal Control with Unstable Model

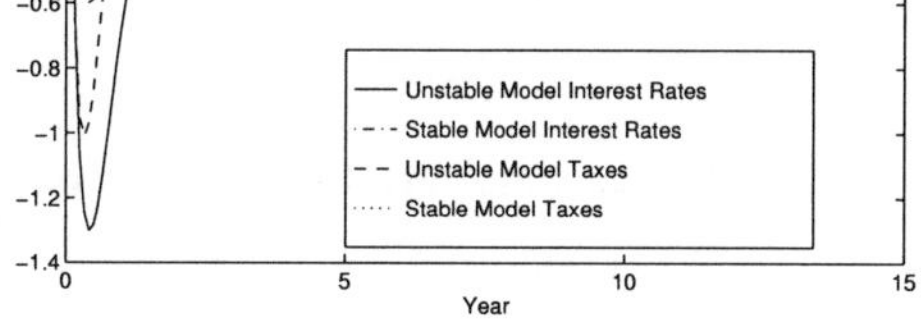

Figure 2.9: Control Generated by Both Stable and Unstable Models

Now consider a situation where the model has more unstable eigenvalues than the saddlepath property requires. This can be achieved by modifying the wealth accumulation equation (Equation 2.49) to become an adjustment process back to the real wealth equilibrium. The equation then becomes

$$dv = [v - \phi_1 s - \phi_2 y + \phi_3(e - p)]\, dt - dp \qquad (2.58)$$

For this modified model (i.e. Equations 2.48, 2.58, 2.50 and 2.51), the eigenvalues of the state equation driving matrix are $\{0.2 \pm 0.4\imath, 0, -0.1\}$. The model now has three eigenvalues with non-negative real parts. In this situation, the observer based method can still generate a set of policies that return the model towards its long run equilibrium. Again, using the same optimal control scenario but with this completely unstable version of the model produces the results presented in Figure 2.8. The results with this version of the model are similar to the original (as presented in Figure 2.5) though GDP is less variable.

Figure 2.9 presents the control policy generated in the illustrations of the stable and unstable models. For interest rates the controls are, essentially, similiar. They are also similar to those produced by the original illustration as presented in Figure 2.6. In all cases a sharp short cut to interest rates are generated. The taxation control differs between the models. The original model and the stable model taxation policies are similar, but with the unstable model this fiscal policy is more vigourous.

These scenarios show how the saddlepath property is not required in this approach. Both the approach of Currie and Levine (1993*b*) and that of McKibbin and Sachs (1991) required the saddlepath property that there be exactly the same number of eigenvalues with non-negative real parts as there are expectations variables. The above scenarios have shown that this is not the situation with the observer based approach. As with the other approaches, the observer based approach assumes that the model is controllable (or, at least, stabilisable). It also reqires that the model be observable (or, at least detectable). But it does not require the saddlepath property. It thus removes a restriction in the design of rational expectations models.

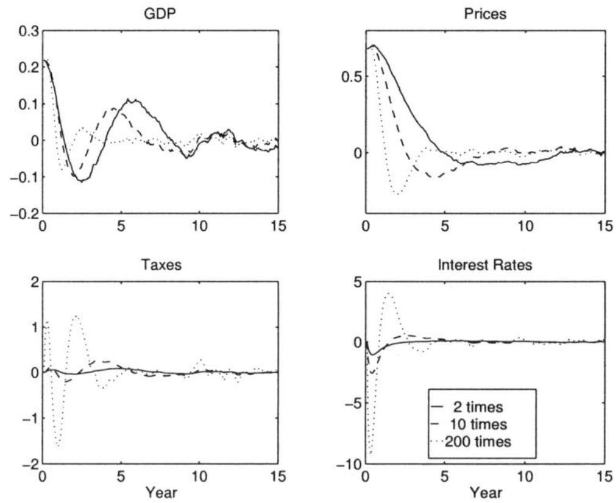

Figure 2.10: Emphasising Output: Implications for Outputs and Controls

These scenarios also indicate one method of policy analysis with these techniques. The three models have substantially different responses in the open loop. Small changes to the original model have produced modified models which generate significantly different multipliers. Yet the closed loop response of each of the models is essentially the same. This is the situation for not only the steady state, but also for the transient response. Further, the control policies generated for each of the models is similar. This is especially the case for interest rates. The scenarios have shown how similar policies are created for changes that have substantial implications for the model.

2.5.4 Policy Analysis

Policy analysis can be illustrated further by considering the implications of different social loss functions on the original model. The trade-offs associated with different weighting on the desired outputs and controls can be examined. For illustrative purposes suppose that policy makers wish to return the model to its long run equilibrium faster than in the original scenario. Can this be achieved, and what are the implications of attempting to achieve it? Suppose that policy makers increase their weighting of deviations of output from its long run

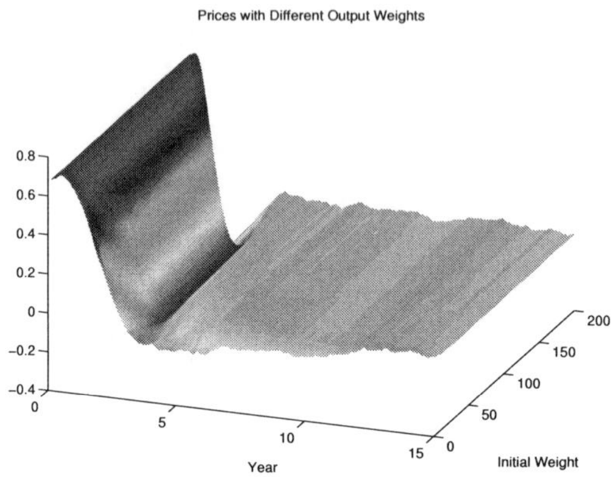

Figure 2.11: Price Level Under Varying Weights

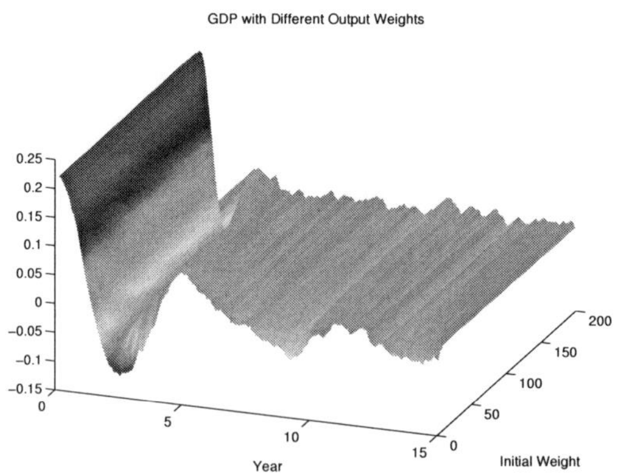

Figure 2.12: GDP Under Varying Weights

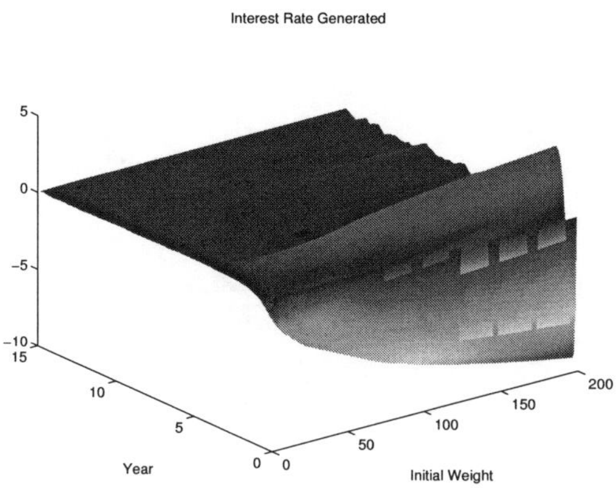

Figure 2.13: Generated Interest Rates Under Varying Weights

equilibrium but leave their weighting of control the same. For this illustration both outputs are weighted equally, but are weighted 2 , 10 and 200 times that of the controls. In the final case then, policy makers are 200 hundred times more concerned about deviations of GDP and prices from their long run equilibrium than they are with autonomous taxes and interest rates. The implications are shown in Figure 2.10, where each output and each control is plotted with the different weights. It can be seen from the Figure that GDP returns to its long run equilibrium faster with the increased empahsis. It is also less variable. The price level also returns to its long run equilibrium faster with increased weights, but with more variability in the transient region. The implications for autonomous taxes are greater. With the increasing emphasis on outputs, greater and more rapid changes are required. With the original weights, taxes only varied slightly from their long run equilibrium, but this is not the case with the increased weights. The implications for interest rates are similiar, but the magnitude of the changes is much larger. Such changes to interest rates would be impractical.

Another view of this scenario is given in Figures 2.11 to 2.14. Here there is a gradual change in the emphasis on the outputs from twice to 200 times the controls. Again, the outputs are weighted equally. The trade-offs can be seen from the Figures. These Figures

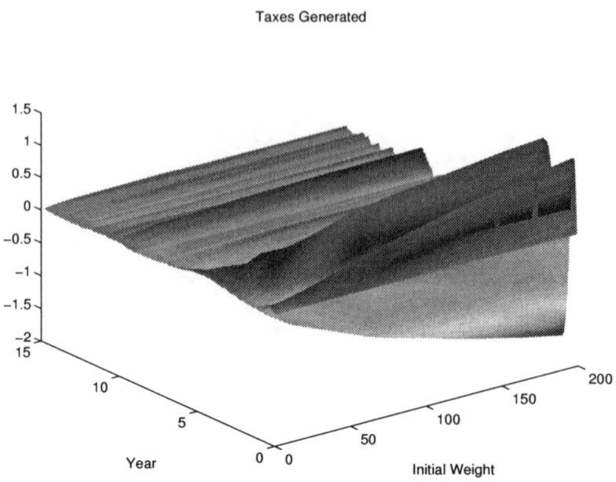

Figure 2.14: Generated Taxes Under Varying Weights

highlight an advantage of the use of optimal control exercises in policy analysis. The model makes explicit the economic assumptions and theory being used. Then, given the model, the trade-offs between various model outputs and inputs can be examined by manipulating the objective function. If it is not possible to achieve the objectives then this will be clearly seen. The economic reasonableness of the results can also be examined. The final result may well be a better economic model.

2.5.5 Normative Rule

Consider the observer based approach with a normative fixed rule of the type Currie and Levine (1993*b*) desire. Here a time invariant feedback rule is used with the observer states. The rule was developed above (on page 34). For this illustration, the same scenario as Currie and Levine (1993*a*) is used with the objective of returning the model outputs to their long run equilibrium; outputs weighted at twice that of controls and no cross product weighting in the social loss function. Figure 2.15 presents the output Gross Domestic Product from the system and the observer, and shows how the observer learns the system's GDP and forces it to return towards its long run equilibrium. Figure 2.16 presents the price level from the

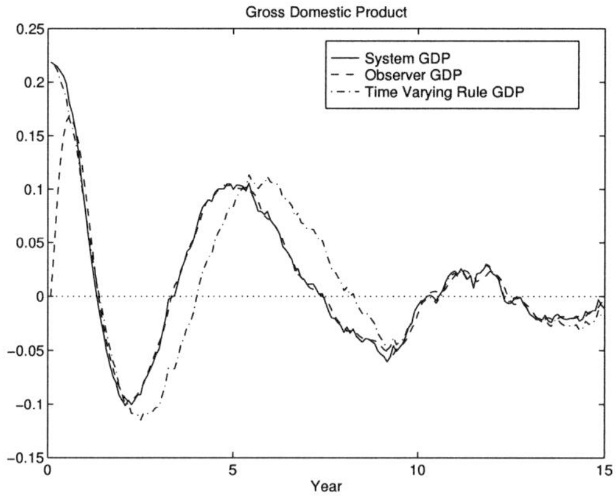

Figure 2.15: GDP Under Fixed and Time Varying Rules

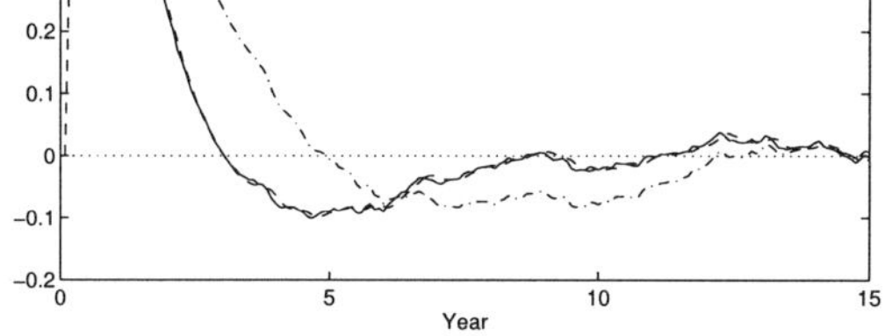

Figure 2.16: Prices Under Fixed and Time Varying Rules

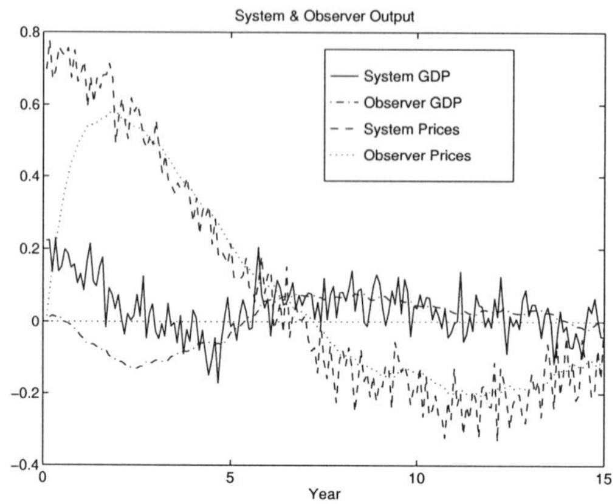

Figure 2.17: Outputs Under Optimal Learning

system and the observer. It also shows that the observer can learn the system's price level and generate control policies that return the price level to its long run equilibrium.

For a comparison between the time invariant and time varying rules, Figures 2.15 and 2.16 plot the results from the time varying rule. In both cases it is the output from the system under the time varying rule that is plotted. It can be seen that the fixed and time varying rules produce very similar results, with the fixed rule being marginally slower at returning the model to its long run equilibrium.

2.5.6 Optimal Learning

The next illustration is of Linear Quadratic Gaussian control. In this illustration, the learning gain is chosen so that it minimises the effects of the known white, gaussian noise in the system. These are the demand and supply shocks. The development of the control rule is given above (on page 36).

Linear Quadratic Gaussian control imposes extra requirements on the model. These requirements are not met by the model (Equations 2.48, 2.49, 2.50, and 2.51), so, for this illustration, the model is modified. The modified model includes noise on the expected exchange rate equation, implying that it is subject to demand shocks. This also illustrates the expectations 'jumping'. The expected exchange rate then becomes

$$de^e = r\,dt + dv_1. \tag{2.59}$$

The scenario for this illustration is similar to those above. The system is given random initial conditions (between zero and one); the observer is assumed to be at its long run equilibrium; the desired outputs and controls are the long run equilibrium ($\mathbf{y_d} = \mathbf{u_d} = \mathbf{0}, \forall t \in [t_0, t_f]$); the outputs are initially weighted at two ($Q(0) = [[2,0]^T, [0,2]^T]$), and the controls at one ($R(0) = [[1,0]^T, [0,1]^T]$); and, the discount rate is 5% ($\rho = 0.05$). The demand and supply shocks each have a variance of 0.1 ($\sigma_1^2 = \sigma_2^2 = 0.1$), and the measurement noise has a variance of 0.05 for each output ($\sigma_3^2 = \sigma_4^2 = 0.05$).

The output from this scenario is plotted in Figure 2.17. From the Figure it can be seen that the system is returned towards its long run equilibrium. The additional noise can also be seen.

2.5.7 Robustness to Parameter Perturbation

The above simulations have shown that the observer can learn the expectations states (and, in fact, all states) and still generate an optimal set of controls. It is instructive to consider if this is still the case when the parameters of the system are perturbed. If a similar set of control policies are generated when the observer and model parameters differ, then greater confidence can be placed in these policies. Further, this situation examines the scenario when the observer is not a perfect representation of the 'real world'. It could be considered as using a weaker form of the rational expectations hypothesis so that economic actors do not have complete knowledge of the parameters of the model. Finally, it examines the situation where the observer has to also learn some of the structure of the model.

A simple consideration of parameter robustness can be seen by perturbing the state equation driving matrix and using the perturbed matrix for the system. It is this matrix that determines, in the open loop, the transient dynamics of the model. Any perturbation will improve the condition of the matrix if it removes the row of zeros. But, the observer will now have to cope with being different from the system it is attempting to control.

Perturbing the model in this way could be considered as allowing the model to have multiplicative disturbances as well as the additive disturbances. The demand and supply shocks could now be considered as not simply additive noise, but to have a multiplicative impact on the model.

A number of experiments are undertaken where the system state equation driving matrix is perturbed through the addition of normally distributed random disturbances of known variance. The state equation driving matrix of the system (Equations 2.20 and 2.21) is perturbed (to A_p) by the addition of white gaussian noise to each element. Perturbing the model in this way will not only alter the parameter values, but will alter the structural form of the model. The theoretical development of the observer approach to control assumes the observer and the system are identical. Thus these experiments test the degree to which this assumption may be relaxed.

For each experiment, the state equation driving matrix is perturbed 500 times with a simulation being run for each of the 500 perturbations. The results given are for the mean of these simulations. Four perturbation variance of 0.001, 0.01, 0.1 and 1 are used. Further, simulations are undertaken for both the eigenvalue placement and the Kalman Filter methods of determining the learning gain.

The simulation scenario is the same as that used in Section 2.5.6; that is, the Linear Quadratic Gaussian illustration with both state and measurement noise, the modified expected exchange rate equation (Equation 2.59) and time varying objective function. For the eigenvalue placement, the learning gain, L, was chosen such that the eigenvalues of $[A - LC]$ were equally spaced between -2 and -4. Again, the scenario is similar to that used by Currie and Levine (1993c).

The results of simulations are presented through a set of summary metrics, rather than visually. The first metric is the condition numbers of the system state equation driving

Variance	0	0.001	0.01	0.1	1
$\kappa_2(A_p)$	∞	7225	690	508	93

Table 2.1: Condition Number of Perturbed State Equation Driving Matrix

Gross Domestic Product										
	Eigenvalue Placement					Kalman Filter				
Variance	0	0.001	0.01	0.1	1	0	0.001	0.01	0.1	1
2-norm	1.19	1.19	1.19	1.37	$O(10)^{20}$	1.04	1.04	1.05	1.44	$O(10)^{20}$
RMS	0.09	0.09	0.09	0.10	$O(10)^{20}$	0.08	0.08	0.08	0.11	$O(10)^{20}$
1-norm	12.86	12.86	12.88	14.94	$O(10)^{20}$	11.04	11.04	11.08	15.72	$O(10)^{20}$
∞-norm	0.29	0.29	0.29	0.29	$O(10)^{20}$	0.29	0.29	0.29	0.31	$O(10)^{20}$
Prices										
	Eigenvalue Placement					Kalman Filter				
Variance	0	0.001	0.01	0.1	1	0	0.001	0.01	0.1	1
2-norm	3.49	3.49	3.45	3.76	$O(10)^{20}$	4.60	4.60	4.60	5.35	$O(10)^{20}$
RMS	0.26	0.26	0.26	0.28	$O(10)^{20}$	0.34	0.34	0.34	0.40	$O(10)^{20}$
1-norm	29.11	29.11	29.16	34.07	$O(10)^{20}$	45.03	45.03	45.12	56.84	$O(10)^{20}$
∞-norm	0.79	0.79	0.79	0.80	$O(10)^{20}$	0.79	0.79	0.80	0.85	$O(10)^{20}$

Table 2.2: Output Metrics with A_p Perturbed

matrix. The rest of the metrics measure the error between the output of the system and the desired output of zero. For each of GDP and prices, the errors are measured by the 2-norm, root mean square (RMS), the 1-norm and the ∞-norm.

Table 2.1 presents the condition number (using the 2-norm) of the state equation driving matrix after it has been perturbed. The Table shows that, as expected, the greater the perturbation variance, the smaller the condition number. Recall the model has a row of zeros in the state equation driving matrix (and thus is infinitely ill-conditioned using the 2-norm) so that increasing the perturbation to the matrix is more likely to improve its condition. This can be seen in the Table.

The metrics for the outputs are presented in Table 2.2. The benchmark simulation run, which has no perturbation to parameters, is also presented in the Table. In this case, where the perturbation variance is zero, the results are generated by the identical scenario, for the Kalman Filter, to that in Section 2.5.6 and are the same as the temporal results plotted in Figure 2.17.

It can be seen from the Table that for small perturbations, the observer method works well. All metrics are similar when the observer and system are slightly different. Once the difference between the observer and the system becomes more substantial, on the other hand, the error between the desired and actual outputs become massive. Both the eigenvalue placement method and Kalman Filter produced similar results. As would be expected the ∞-norm is the first time period and prices has more error than GDP.

The main conclusion from these simulations is that the approach is relatively robust to small amounts of parameter uncertainty. Thus if economic actors do not have full information of the model, which is surely the case in practice, then the observer approach is still appropriate.

2.6 Conclusion

This Chapter has presented an observer based approach for the development of optimal control with linear stochastic rational expectations models. The observer is separate from the actual system and observes its behaviour. It has the same basis as the system but includes extensions to learn the actual model's behaviour through output-injection. The control policy is generated by the observer and applied to the system. The main emphasis has been on learning the expectations states. The major advantage of this method is that standard optimal control techniques can be used in conjunction with it. The technique is a standard control technique.

The control policy can be generated by a simple rule which does not include an integrator. This is the time-consistent type of rule for which Currie and Levine (1993*c*) aimed. Further, the approach does not require the saddle path property, as in the approach of Currie and

Levine (1993*c*). The approach of McKibbin and Sachs (1991) requires both the saddle path property and knowledge of terminal conditions. The learning based approach does not require knowledge of terminal conditions.

The observer based approach assumes knowledge of the system by which expectations are formed and learns the states from this system. Under the rational expectations hypothesis, knowledge of this system is given as economic actors have full knowledge of the model and its parameters. This Chapter also considered the implications of a weaker form of this hypothesis where incomplete knowledge is assumed with a (slightly) different observer from the system that generated the expectations.

Currie and Levine's (1993*c*) use of optimal control theory was for the development of normative fixed policy rules. Here the emphasis has been on the use of optimal control theory for policy analysis whereby the trade-offs associated with temporal policy objectives can be seen. The questions have been more concerned with issues such as whether temporal policy objectives are achievable, and what are the implications of attempting to achieve them, rather than on designing specific policy rules to be followed by policy makers.

This Chapter has developed the theoretical basis for the observer based approach. It has shown how expectations states can be learnt from knowledge of the model's output. It has also shown that the process of learning is separate from the process of developing the optimal control. Two methods of determining learning gains have been presented in the eigenvalue placement method and in the optimal learning gain or Kalman Filter.

With the use of a specific model developed by Currie and Levine (1993*c*), the application of the methodology has been illustrated. The learning of expectations states has been illustrated, as has been the development of optimal policies. By simulations it has been illustrated how the eigenvalue placement method is more robust than the Kalman Filter.

2.7 Appendix: The Linear Quadratic Tracking Problem

This Appendix derives the Linear Quadratic Tracking Problem. It takes the simple case, for, from it, the other situations can be developed. The aim is derive the optimal control rule of the type used in this Chapter. The Currie and Levine (1993*c*) analysis can be derived using the same approach.

The problem will be formulated as: minimise the objective function, given by

$$J = \frac{1}{2}\bar{\mathbf{y}}(t_f)^T Q(t_f)\bar{\mathbf{y}}(t_f) + \frac{1}{2}\int_{t_0}^{t_f} (\bar{\mathbf{y}}(t)^T Q(t)\bar{\mathbf{y}}(t) + \bar{\mathbf{u}}(t)^T R(t)\bar{\mathbf{u}}(t))\, dt \tag{2.60}$$

where the bars indicated deviations from desired trajectories so

$$\bar{\mathbf{y}}(t) = \mathbf{y}(t) - \mathbf{y_d}(t) \tag{2.61}$$

$$\bar{\mathbf{u}}(t) = \mathbf{u}(t) - \mathbf{u_d}(t) \tag{2.62}$$

and $\mathbf{y_d}(t)$ and $\mathbf{u_d}(t)$ are given $\forall t$.

Subject to the model, given by

$$\dot{\mathbf{x}}(t) = A(t)\mathbf{x}(t) + B(t)\mathbf{u}(t) \tag{2.63}$$

$$\mathbf{y}(t) = C(t)\mathbf{x}(t) \tag{2.64}$$

with $\mathbf{x}(0)$ given.

This is the deterministic output tracking problem with no direct effect of the control on the output. It has been specified as time varying, but without loss of generality, the time invariant case can be considered. So references to time will only be given in the boundary conditions.

The meaning of the variables and the necessary conditions are given above (on pages 29 and 33).

Define the Hamiltonian

$$\begin{aligned} H(\mathbf{x},\lambda,t) &= \frac{1}{2}(\bar{\mathbf{y}}^T Q \bar{\mathbf{y}} + \bar{\mathbf{u}}^T R \bar{\mathbf{u}}) + \lambda^T (A\mathbf{x} + B\mathbf{u}) && (2.65)\\ &= \frac{1}{2}[(C\mathbf{x} - \mathbf{y_d})^T Q (C\mathbf{x} - \mathbf{y_d}) + (\mathbf{u} - \mathbf{u_d})^T R (\mathbf{u} - \mathbf{u_d})] \\ &\quad + \lambda^T (A\mathbf{x} + B\mathbf{u}) && (2.66) \end{aligned}$$

Now by Pontryagin's minimum principle (Anderson and Moore, 1990), the optimal control and state trajectories must satisfy the following three conditions.

Firstly,

$$\frac{\partial H}{\partial \mathbf{u}} = \mathbf{0} = R(\mathbf{u} - \mathbf{u_d}) + B^T \lambda \tag{2.67}$$

which gives the optimal control

$$\mathbf{u} = -R^{-1} B^T \lambda + \mathbf{u_d} \tag{2.68}$$

Secondly,

$$\frac{\partial H}{\partial \lambda} = \dot{\mathbf{x}} \tag{2.69}$$

which gives the state equation

$$\dot{\mathbf{x}} = A\mathbf{x} + B\mathbf{u} \tag{2.70}$$

with the boundary condition

$$\mathbf{x}(t_0) = \mathbf{x_0} \tag{2.71}$$

Thirdly,

$$-\frac{\partial H}{\partial \mathbf{x}} = \dot{\lambda} \tag{2.72}$$

which gives the co-state equation

$$\dot{\lambda} = -[C^T Q (C\mathbf{x} - \mathbf{y_d}) + A^T \lambda] \tag{2.73}$$

with the boundary condition

$$\lambda(t_f) = C(t_f)Q(t_f)[C(t_f)\mathbf{x}(t_f) - \mathbf{y_d}(t_f)] \tag{2.74}$$

This produces a coupled set of linear differential equations which form a two point boundary value problem with mixed boundary conditions.

Combining the optimal control into the state equation, it can be written as the Hamiltonian Equation

$$\begin{bmatrix} \dot{\mathbf{x}} \\ \dot{\lambda} \end{bmatrix} = \begin{bmatrix} A & -BR^{-1}B^T \\ -C^TQC & -A \end{bmatrix} \begin{bmatrix} \mathbf{x} \\ \lambda \end{bmatrix} + \begin{bmatrix} B\mathbf{u_d} \\ C^TQ\mathbf{y_d} \end{bmatrix} \tag{2.75}$$

The problem can be solved analytically by the sweep method. Try for a solution. Let

$$\lambda = S\mathbf{x} + \mathbf{g} \tag{2.76}$$

then

$$\dot{\lambda} = S\dot{\mathbf{x}} + \dot{S}\mathbf{x} + \dot{\mathbf{g}} \tag{2.77}$$

Substitute the optimal control (Equation 2.68) into the state equation (Equation 2.70) and the resulting equation together with the co-state equation (Equation 2.73) into Equation 2.77, giving

$$-[C^TQ(C\mathbf{x} - \mathbf{y_d}) + A^T\lambda] = S(A\mathbf{x} - BR^{-1}B^T\lambda + B\mathbf{u_d}) + \dot{S}\mathbf{x} + \dot{\mathbf{g}} \tag{2.78}$$

which can be rearranged to

$$\dot{S}\mathbf{x} + \dot{\mathbf{g}} = (-SA - A^TS + SBR^{-1}B^TS - C^TQC)\mathbf{x} + (SBR^{-1}B^T) - A^T)\mathbf{g} \\ + C^TQ\mathbf{y_d} - SB\mathbf{u_d} \tag{2.79}$$

Then, by equating coefficients, the Riccati Equation

$$\dot{S} = -SA - A^TS + SBR^{-1}B^TS - C^TQC \tag{2.80}$$

and the predictor equation

$$\dot{\mathbf{g}} = (SBR^{-1}B^T - A^T)\mathbf{g} + C^T Q\mathbf{y_d} - SB\mathbf{u_d} \tag{2.81}$$

are obtained.

For the boundary conditions,

$$\begin{aligned} S(t_f)\mathbf{x}(t_f) + \mathbf{g}(t_f) &= \lambda(t_f) && (2.82) \\ &= C(t_f)^T Q(t_f)\mathbf{x}(t_f) - C(t_f)^T Q(t_f)\mathbf{y_d}(t_f) && (2.83) \end{aligned}$$

and by equating coefficients

$$\begin{aligned} S(t_f) &= C(t_f)^T Q(t_f) && (2.84) \\ \mathbf{g}(t_f) &= -C(t_f)^T Q(t_f)\mathbf{y_d}(t_f) && (2.85) \end{aligned}$$

The optimal control rule can then be rearranged so that it is not necessary to calculate the co-states. This rule is then given by

$$\mathbf{u} = -R^{-1}B^T(S\mathbf{x} + \mathbf{g}) + \mathbf{u_d} \tag{2.86}$$

For the implementation, the Riccati and predictor equations can be solved before the control experiment. The control experiment is run out forwards in time and in this way the control can be adapted for disturbances, such as expectations 'jumping' to 'news'.

Chapter 3

LOW ORDER DYNAMIC MACROECONOMETRIC MODELS

3.1 Introduction

In this Chapter a suite of dynamic macroeconometric models are built. The emphasis is on discrete time, low order, macroeconomic models. The models developed in this Chapter are used in later Chapters of this book for illustrative purposes and to form the basis of observers for the large econometric model developed in Chapter 7. These models form a suite of linear, nonlinear, discrete time and continuous time low order models.

Three classes of models are developed in this Chapter. The first is based on on the Multiplier Accelerator theory (Section 3.2); the second is based on IS-LM theory (Section 3.3); and, the third (in Section 3.4) is based on statistical time series relationships rather than on economic structure.

As compared to many engineering control models, dynamic discrete time macroeconomic models usually have the control to be contemporaneous with the output. The models developed here include contemporaneous control. Exogenous variables and level constants are also important in economic models, and they, too, are explicitly considered in the models developed in this Chapter.

The models developed in this Chapter are transformed into a state space form. This state space form differs from that usually used with economic models in that it includes an output equation. It is shown that the inclusion of an output equation gives a structure that allows contemporaneous control to be incorporated into the state space form. The state space form also differs from the standard engineering control theory in that it includes level constants and exogenous variables.

The models are estimated using Quarterly Australian data, using the standard econometric approach of using the models in structural form. Both the Ordinary Least Squares and Three Stage Least Squares estimation routines are used. The models are also estimated in state space form, where the emphasis is on obtaining a good fit to the historical data for the model outputs, when the model is run out from initial conditions.

3.2 The Multiplier Accelerator Model

The Multiplier Accelerator model is a simple, standard macroeconomic model. The model is frequently used as an example in macroeconomic dynamics (e.g. Pindyck and Rubinfield, 1991). It is concerned with aggregate demand in the economy; and, in its simplest form, breaks this into a consumption function, an investment function, and the national income identity.

A linear version of the model may be written in structural form as

$$C_t = \alpha_1 + \alpha_2 Y_{t-1} \tag{3.1}$$

$$I_t = \beta_1 + \beta_2 (Y_{t-1} - Y_{t-2}) \tag{3.2}$$

$$Y_t = C_t + I_t + G_t \tag{3.3}$$

Equation 3.1 is the consumption function, and C is private consumption expenditure, and Y is Gross National Expenditure (GNE). Equation 3.2 is the investment function, and I is private investment expenditure. Equation 3.3 is the national income identity, with G being general government expenditure. Here t refers to discrete (usually Quarterly) time periods; and the variables with Greek letters are the parameters of the model. The model is usually used with policy makers having influence over the policy instrument (or control) of government expenditure and desiring to influence the policy target of GNE (the output).

The model may be written in the state space form of

$$\mathbf{x}_{t+1} = A\mathbf{x}_t + B\mathbf{u}_t + E \tag{3.4}$$

$$\mathbf{y}_t = C\mathbf{x}_t + D\mathbf{u}_t \tag{3.5}$$

with

$$\begin{array}{llll} \mathbf{y}_t = Y_t & \mathbf{u}_t = G_t & C = \begin{bmatrix} 1 & 1 \end{bmatrix} & D = 1 \\ \mathbf{x}_t = \begin{bmatrix} C_t \\ I_t \end{bmatrix} & A = \begin{bmatrix} \alpha_2 & \alpha_2 \\ \beta_2 - \gamma & \beta_2 \end{bmatrix} & B = \begin{bmatrix} \alpha_2 \\ \beta_2 \end{bmatrix} & E = \begin{bmatrix} \alpha_1 \\ \beta_1 + \alpha_1\gamma \end{bmatrix} \\ \gamma = \dfrac{\beta_2}{\alpha_2} & & & \end{array} \tag{3.6}$$

This expresses the model as a single input, single output, second order, discrete, state space model. It produces a simple model for use in this book. A detailed analysis of this model, both theoretically and econometrically, and its use in control exercises, can be found in Herbert (1998).

This state form of Equations 3.4 and 3.5 compares to that of Kendrick (1981), where the model is also used in control exercises. In that case, no output equation (Equation 3.5) is used, and the concern is with controlling consumption and investment. The use of the output equation allows for the model to be used as usual with the emphasis on GNE. It also allows for the simple expression of the national income identity with its contemporaneous control. Kendrick (1981) effectively sidesteps the contemporaneous control by lagging government expenditure one time period and calling this government obligations.

This state space form also differs from the standard control engineering state space form in that it includes level constants. The usual control approach is to define the system as operating around some level, so that there is no need for level constants. The level constants are particularity relevant to this work where linear models form the basis of observers for nonlinear models. The level constants do not effect the behaviour of the transient dynamics of the system but only its magnitude, so, given the incorporation of the levels, the standard control algorithms still apply.

One of the advantages of the state space form of Equations 3.4 and 3.5 is that it can easily be extended to include exogenous variables. The economics literature considers exogenous variables important, whereas their explicit inclusion in control engineering models has become more important with the advent of robust control (Doyle, Glover, Khargonekar and Francis, 1989).

The inclusion can be done as follows: suppose the system is influenced by a vector of length r exogenous variables, then define $\mathbf{z}$ as the a vector of length $r+1$ being augmented with unity as, say, the first element. The E (in Equation 3.4) becomes an $(r+1) \times (r+1)$ matrix of parameters, and $\mathbf{z}$ can now be included in the state equation to allow for both level constants and exogenous variables (Kendrick and Amman, 1992). The state space system then becomes

$$\mathbf{x}_{t+1} = A\mathbf{x}_t + B\mathbf{u}_t + E\mathbf{z}_t \tag{3.7}$$

$$\mathbf{y}_t = C\mathbf{x}_t + D\mathbf{u}_t \tag{3.8}$$

3.3 IS-LM Models

The IS-LM approach generates a model structure representing basic Keynesian economics and attempts to capture the key interactions between commodities and money in the macroeconomy. The structure allows for the development of a variety of economic models, and it is used in this Section to build a number of low order macroeconometric models. A number of models are also built by extending the IS-LM framework to include prices.

3.3.1 Structural Form

In its simple form, the IS-LM framework has an equation representing equilibrium in the *commodity market* (or goods market), and another representing equilibrium in the *money market* (or assets market). The IS (Investment equals Savings) function gives the loci of equilibrium in the commodity market, and the LM (liquidity preference equals money) function gives the loci of equilibrium in the money market. Usually the functional forms of the IS and LM functions are not specified.

A standard form of the model, as presented by Blanchard and Fischer (1989), has an IS function given by Equation 3.9, where Y is real economic output (GNE), i is the real interest rate, and F is some fiscal policy instrument (control).

$$Y = f(i, Y, F), \qquad \frac{\partial f}{\partial i} < 0, \quad 1 > \frac{\partial f}{\partial Y} > 0, \quad \frac{\partial f}{\partial F} > 0 \tag{3.9}$$

The LM function is given by Equation 3.10 where r is the nominal interest rate, M is the demand for money and P is the price level.

$$\frac{M}{P} = g(r, Y), \qquad \frac{\partial g}{\partial i} < 0, \quad \frac{\partial g}{\partial Y} > 0 \tag{3.10}$$

The model has its limitations. It ignores the supply side of the economy; and, variables that effect the demand for commodities, may also effect their supply. Prices are assumed fixed (or sticky), so the model is only appropriate for short-run analysis. Further, the modelling of the demand for commodities is often considered to be simplistic in that it ignores, for example, wealth and expectations. Despite its shortcomings, Blanchard and Fischer (1989, p. 531), argue that the IS-LM model is an appropriate model for transient output dynamics.

The model is frequently used as the basis for more elaborate economic models, with, for example, it being augmented with wealth accumulation. It is often used as the basis of large macroeconometric models after some functional form (often linear) is specified. In the usual convention of macroeconometric models, the equations are normally expressed in a discrete form and frequently also include a complex lag structure. This is the case with the large econometric model considered in Chapter 7.

The model is often expressed in a linear form (e.g. Crouch, 1972; Parkin and Bade, 1992), and the IS-LM model developed in this book will have its basis in the linear model. It will, though, remain consistent with the general form given by Equations 3.9 and 3.10.

For the linear form, the *commodity market* is given by

$$Y_s = Y_d \tag{3.11}$$

$$Y_d = C + I + G \tag{3.12}$$

$$C = \alpha_0 + \alpha_1 (Y_s - T) - \alpha_2 r \tag{3.13}$$

$$I = \beta_0 + \beta_1 Y_s - \beta_2 r \tag{3.14}$$

$$T = \gamma_0 + \gamma_1 Y_s \tag{3.15}$$

Equation 3.11 is the equilibrium condition for the commodity market with aggregate supply (Y_s) and aggregate demand (Y_d) being equal. Equation 3.12 is the national income identity which defines planned aggregate demand as the sum off consumption expenditure (C), investment expenditure (I) and (the exogenous) government expenditure (G). Government expenditure has been specified as the fiscal policy control, when policy analysis is considered. Equation 3.13 is the consumption function, with consumption being determined by disposable income and interest rates. Equation 3.14 is the investment function with investment determined by income and interest rates. The final equation, Equation 3.15, relates tax revenue (T) to income.

The *money market* is given by

$$M_s = M_d \tag{3.16}$$

$$\frac{M_d}{P} = \delta_0 + \delta_1 Y_s - \delta_2 r \tag{3.17}$$

Equation 3.16 gives the equilibrium condition in the money market with the demand for money (M_d) equal to the (exogenous) supply of money (M_s). The demand for real money balances is given by Equation 3.17 where it is positively related to aggregate supply (national income), and negatively related to interest rates. In this model, the price level (P) is exogenous.

3.3.2 Adjustment Process Form

This model is a steady state model, describing the economy at equilibrium; and, it is frequently used in this form for comparative static analysis. For the purposes here, the model is converted into an adjustment process form where it explains the movement of the economy to the equilibrium. Firstly, a reduced form of the commodity market can be developed by solving for aggregate demand. Leaving interest rates as a linkage to the money market, and the government expenditure as exogenous, then from Equations 3.11 to 3.15

$$Y_d = (\alpha_0 + \beta_0 - \alpha_1\gamma_0) - (\alpha_2 + \beta_2)r + (\alpha_1 - \beta_1 + \alpha_1\gamma_1)Y_s + G \tag{3.18}$$

Equilibrium in the commodity market is attained by aggregate demand and aggregate supply adjusting until they are equal. This adjustment process can be modelled as

$$\frac{dY}{dt} = \mu_1(Y_d - Y) \tag{3.19}$$

$$= \mu_1(\alpha_0 + \beta_0 - \alpha_1\gamma_0) + \mu_1(\alpha_1 - \beta_1 + \alpha_1\gamma_1 - 1)Y - \mu_1(\alpha_2 + \beta_2)r + \mu_1 G \tag{3.20}$$

This assumes that the economy's income (Gross National Expenditure) is given by aggregate supply so that Y is defined as Y_s. It also allows for the adjustment process to be sluggish with a mean lag of $\frac{1}{\mu_1}$, $(\mu_1 > 0)$.

Similarly, in the money market interest rates adjust according to the demand and supply of real money balances. This is, again, modelled as the sluggish process given in Equation 3.21, with, again, $\mu_2 > 0$.

$$\frac{dr}{dt} = \mu_2\left(\frac{M_d}{P} - \frac{M_s}{P}\right) \tag{3.21}$$

$$= \mu_2\delta_0 + \mu_2\delta_1 Y - \mu_2\delta_2 r - \mu_2\left(\frac{M_s}{P}\right) \tag{3.22}$$

Notice that the Equation 3.22 is linear in national income, interest rates and the real money supply; for, in this model, both the nominal money supply (M_s) and the price level (P) are exogenous.

This produces a linear adjustment model as given by Equations 3.20 and 3.22 that describes the economy's movement between two equilibrium positions. It can be written as the state equation

$$\begin{bmatrix} \frac{dY(t)}{dt} \\ \frac{dr(t)}{dt} \end{bmatrix} = \begin{bmatrix} \mu_1(\alpha_1 - \beta_1 + \alpha_1\gamma_1 - 1) & -\mu_1(\alpha_2 + \beta_2) \\ \mu_2\delta_1 & -\mu_2\delta_2 \end{bmatrix} \begin{bmatrix} Y(t) \\ r(t) \end{bmatrix} + \begin{bmatrix} \mu_1 & 0 \\ 0 & -\mu_2 \end{bmatrix} \begin{bmatrix} G(t) \\ \frac{M_s(t)}{P(t)} \end{bmatrix} + \begin{bmatrix} \mu_1(\alpha_0 + \beta_0 - \alpha_1\gamma_0) \\ \mu_2\delta_0 \end{bmatrix} \tag{3.23}$$

This model can be written more compactly as

$$\dot{\mathbf{x}}(t) = A\mathbf{x}(t) + B\mathbf{u}(t) + E \tag{3.24}$$

The model has states ($\mathbf{x}$) (and outputs) of national income and interest rates, and controls ($\mathbf{u}$) of government expenditure and the real money supply.

By defining an output matrix (C) as the (2×2) identity matrix and the contemporaneous control matrix (D) as the (2×2) zero matrix, the model forms a low order continuous time state space model.

3.3.3 Adding Prices

The model can be extended to include prices. One method is to extend the model by the introduction of a Phillips curve which links inflation and unemployment. But, monetary variables are also considered to be an important determinant of the price level; and so, too, are the inflationary expectations of forward looking economic actors.

An expectations augmented Phillips curve specifies that inflation is a function of unemployment and inflationary expectations. A standard function of this type is

$$\frac{\frac{dP}{dt}}{P} = \frac{d\log P}{dt} = h(U, \frac{dU}{dt}) + \frac{d\log P^e}{dt} \qquad \frac{dh}{dU} \leq 0 \quad \frac{dh}{\frac{du}{dt}} \leq 0 \tag{3.25}$$

where U is the level of unemployment and the superscript e refers to expectations of forward looking economic actors. This specifies that inflation is a function of the unemployment level, and its change, plus the expected inflation rate.

The functional form is usually specified as a relative deviation from some 'natural' rate of unemployment, so the Phillips curve is usually written in the form presented in Equation 3.26. Here u is the unemployment rate and u^n is the natural rate of unemployment or Non-Accelerating Inflation Rate of Unemployment (NAIRU). That is,

$$\frac{d \log P}{dt} = \frac{u - u^n}{u^n} + \frac{d \log P^e}{dt} \tag{3.26}$$

In the short run, the key variable for unemployment is the level of national income, so the Phillips curve is often written as a function of the level of national income adjusting back to its full employment level (Turnovsky, 1995, 33). Ignoring, for now, the expectations component, the Phillips curve can then be written as

$$\frac{d \log P}{dt} = \xi(Y^f - Y) \tag{3.27}$$

where Y^f is the full employment level of national income. For short run adjustments, Equation 3.27 can be approximated by a linear relationship of the level of national income.

The money supply is also considered to be an important determinant of prices. It frequently enters the model through either levels or as changes in levels. So, including the expectations component, the money supply component and the level of national income component as a linear combination, gives an inflation equation of

$$\frac{d \log P}{dt} = \tau_0 - \tau_1 Y + \tau_2 M_s + \tau_3 \frac{d \log P^e}{dt} \tag{3.28}$$

Adding this inflation equation to the adjustment equations for national income and interest rates, creates the model made up of Equations 3.20, 3.22 and 3.28. This forms a suite of low order, nonlinear models depending on, for example, whether inflationary expectations are to be considered in the prices equation. Of course, using different forms of the IS (Equation 3.9), LM (Equation 3.10) and Phillips (Equations 3.25) curves generates an even broader suite of models.

More models can be formed by using linearised versions. Following the usual economic practice, the model can be linearised by measuring variables in logarithms. Without forward looking inflationary expectations, this produces the model given[1] by

$$\begin{aligned}\frac{dY}{dt} &= \mu_1(\alpha_0+\beta_0-\alpha_1\gamma_0)+\mu_1(\alpha_1-\beta_1+\alpha_1\gamma_1-1)Y \\ &\quad -\mu_1(\alpha_2+\beta_2)r+\mu_1 G && (3.29)\\ \frac{dr}{dt} &= \mu_2\delta_0+\mu_2\delta_1 Y-\mu_2\delta_2 r+\mu_2 P-\mu_2 M_s && (3.30)\\ \frac{dP}{dt} &= \tau_0-\tau_1 Y+\tau_2 M_s && (3.31)\end{aligned}$$

With such a model, national income and prices would normally be considered the outputs, so this leads to a two input, two output, third order model. This model may be written in the state space form of

$$\begin{aligned}\dot{\mathbf{x}}(t) &= A\mathbf{x}(t)+B\mathbf{u}(t)+E && (3.32)\\ \mathbf{y}(t) &= C\mathbf{x}(t)+D\mathbf{u}(t) && (3.33)\end{aligned}$$

with system matrices

$$A=\begin{bmatrix}\mu_1(\alpha_1-\beta_1+\alpha_1\gamma_1-1) & -\mu_1(\alpha_2+\beta_2) & 0\\ \mu_2\delta_1 & -\mu_2\delta_2 & \mu_2\\ -\tau_1 & 0 & 0\end{bmatrix}\quad B=\begin{bmatrix}\mu_1 & 0\\ 0 & -\mu_2\\ 0 & \tau_2\end{bmatrix}$$

$$C=\begin{bmatrix}1 & 0 & 0\\ 0 & 0 & 1\end{bmatrix}\quad D=\begin{bmatrix}0 & 0\\ 0 & 0\end{bmatrix}$$

$$E=\begin{bmatrix}\mu_1(\alpha_0+\beta_0-\alpha_1\gamma_0)\\ \mu_2\delta_0\\ \tau_2\end{bmatrix}$$

and states of national income, interest rates and prices, $\mathbf{x}(t)=[Y(t),r(t),P(t)]^T$; controls of government expenditure and the money supply, $\mathbf{u}(t)=[G(t),M_s(t)]^T$; and outputs of national income and prices, $\mathbf{y}(t)=[Y(t),P(t)]^T$.

[1] For ease of notation, the same variable and parameter names will be used.

3.3.4 Estimation of the Model

Economic data series are measured at discrete intervals, usually Quarterly. The usual econometric method is to use a discrete form of the model. This is derived by using a forward difference (first order Taylor series) approximation for the derivative. The time step is one (Quarter) due to the data collection frequency, and with the general feeling that truncation is not the major source of error. This approach is followed here. Also, here, a central difference (or three point) approximation of the derivative is used (Burden and Faires, 1985). This reduces the truncation error, and has the advantage of introducing a further lag. Long lag structures are often used in econometric models. Lag structure difference equations are further considered in Section 3.4.

An important difference between econometric modelling and control engineering modelling is the problem of obtaining data. Most economics data series are difficult to collect and are expensive. For many control engineering applications data is relatively easy to obtain, so the usual practice is to collect data from the plant, estimate parameters on this set of data, and verify the model on another set of data from the plant. Because of the difficulties of obtaining data, this is not common in econometrics. The historical data used in this book is limited to Quarterly Australian data for slightly over a 15 year time horizon.

One of the purposes of these models is to act as the basis of observers for control experiments with the large econometric model considered in Chapter 7. The data series definitions match that model. Nominal Gross National Expenditure (GNEz) is used for Y; the Thirty Day Bank Bill (RCST) is used for r; and the Price of the Domestic Good (PYD) is used for P. The Nominal Money Supply (M) is used for the monetary policy control, M_s, and, as a fiscal policy control, the Target Ratio of Public Debt to Gross Domestic Product (RPUBDE) is used. The fiscal policy control differs from that of government expenditure, but is in line with the basic formulation of the IS function (Equation 3.9). The reason for not using government expenditure is that it is not exogenous in the large econometric model. RPUBDE acts as a fiscal control by effectively altering income taxation policy. In some cases, General Government Purchases of Consumption Goods (GGCO) is used as it is exogenous to the large econometric model and is a component of government expenditure.

A major problem is that it is unlikely that any of the data series (except interest rates) will be stationary. While this will not effect the estimated parameters, it will mean that the error terms cannot be considered as white gaussian noise unless the non-stationary series are cointegrated. Thus statistics such as standard errors of the parameters may be biased. This may not be as great a problem for using the model as an observer, where the aim of the model is to behave like the system being observed, rather than test an hypothesis of economic relationships.

3.3.5 Three State Model

The forward difference model can then be written, with all parameters as positive, as

$$Y_{t+1} = \theta_{10} + \theta_{11} Y_t + \theta_{12} r_t + \theta_{13} F_t + v_{1,t} \tag{3.34}$$

$$r_{t+1} = \theta_{20} + \theta_{21} Y_t + \theta_{22} r_t + \theta_{23} M_t + \theta_{24} P_t + v_{2,t} \tag{3.35}$$

$$P_{t+1} = \theta_{30} + \theta_{31} Y_t + \theta_{32} P_t + \theta_{33} M_t + v_{3,t} \tag{3.36}$$

Here F is the fiscal control and M the money supply control; t refers to the discrete (Quarterly) time period; and v_1, v_2 and v_3 are error terms.

This discrete model can be written in a state space form as

$$\mathbf{x}_{t+1} = A\mathbf{x}_t + B\mathbf{u}_t + E + H\mathbf{v}_t \tag{3.37}$$

$$\mathbf{y}_t = C\mathbf{x}_t + D\mathbf{u}_t \tag{3.38}$$

with, again, the states, $\mathbf{x}_t = [Y_t, r_t, P_t]^T$; the controls $\mathbf{u}_t = [F_t, M_t]^T$; the outputs $\mathbf{y}_t = [Y_t, r_t]^T$; and the noise $\mathbf{u}_t = [v_{1,t}, v_{2,t}, v_{3,t}]^T$. The system matrices are

$$A = \begin{bmatrix} \theta_{11} & \theta_{12} & 0 \\ \theta_{21} & \theta_{22} & \theta_{24} \\ \theta_{31} & 0 & \theta_{32} \end{bmatrix} \quad B = \begin{bmatrix} \theta_{13} & 0 \\ 0 & \theta_{23} \\ 0 & \theta_{33} \end{bmatrix} \quad E = \begin{bmatrix} \theta_{10} \\ \theta_{20} \\ \theta_{30} \end{bmatrix} \quad H = \begin{bmatrix} 1 & 0 & 0 \\ 0 & 1 & 0 \\ 0 & 0 & 1 \end{bmatrix}$$

$$C = \begin{bmatrix} 1 & 0 & 0 \\ 0 & 0 & 1 \end{bmatrix} \quad D = \begin{bmatrix} 0 & 0 \\ 0 & 0 \end{bmatrix}$$

OLS Estimates for Equation 3.34 (GNE)			
Parameter Name	Estimated Value	Standard Error	t-Value (53 d.f)
θ_{10}	0.12196	0.0474	2.5728
θ_{11}	0.99218	0.0057	173.61
θ_{12}	-0.00672	0.0108	-0.62046
θ_{13}	0.02692	0.0159	1.6973
$R^2 = 0.9991$		$DW = 1.7642$	

Table 3.1: Ordinary Least Squares Estimates for National Income

There remains the issue of the parameter estimation techniques. Here three least squares estimation procedures are used. The first procedure is the ordinary least squares (OLS) estimation of each of the individual structural equations for national income, interest rates and prices (Equations 3.34, 3.35 and 3.36). The results from this procedure are given in Tables 3.1, 3.2 and 3.3. The second procedure uses the joint estimation of the individual structural equations as a Seemingly Unrelated Regression (SUR). The results of this estimation are given in Table 3.4. Both of these procedures are standard econometric procedures and a standard econometrics package was used to generate the estimates. (See Appendix A on page 304).

The final procedure estimates the model in state space form with the emphasis on the outputs. In this procedure the objective function is the sum of the 2-norm of the difference between each output generated by the state space model when run out from initial conditions (Equations 3.37 and 3.38) and the historical data for that output. The procedure minimises the objective function by varying the parameters. This is a nonlinear problem and an optimisation routine (a Quasi Newton method) is used to find the set of parameters with the initial guess being the ordinary least squares estimates. This is a nonlinear form of least squares. The results of this estimation are given in Table 3.5.

The ordinary least squares estimates for national income, as presented in Table 3.1, are as would be expected. The signs of the parameters are as specified by the model. At standard levels of confidence, the data fits the equation well; the interest rate component is weak, but

OLS Estimates for Equation 3.35 (Interest Rates)			
Parameter Name	Estimated Value	Standard Error	t-Value (52 d.f)
θ_{20}	-10.073	6.4250	-1.5678
θ_{21}	2.5425	1.1944	2.1286
θ_{22}	0.48672	0.1243	3.9157
θ_{23}	-1.8278	0.9214	-1.9837
θ_{24}	-0.77893	0.73904	-1.0540
$R^2 = 0.6416$		$DW = 1.7881$	

Table 3.2: Ordinary Least Squares Estimates for Interest Rates

OLS Estimates for Equation 3.36 (Prices)			
Parameter Name	Estimated Value	Standard Error	t-Value (53 d.f)
θ_{30}	-0.72999	0.3447	-2.1177
θ_{31}	0.12425	0.0591	2.1043
θ_{32}	0.90780	0.0405	22.441
θ_{33}	-0.06819	0.0442	-1.5443
$R^2 = 0.9994$		$DW = 2.2627$	

Table 3.3: Ordinary Least Squares Estimates for Prices

SUR Estimates of Parameters and Standard Errors for the Structural Model (Equations 3.34, 3.35 & 3.36)					
Parameter	Value	Parameter	Value	Parameter	Value
θ_{10}	0.12065	θ_{20}	-11.618	θ_{30}	-0.74922
	(0.04569)		(6.096)		(0.33226)
θ_{11}	0.99238	θ_{21}	2.9459	θ_{31}	0.12862
	(0.00551)		(1.1268)		(0.05691)
θ_{12}	-0.0069577	θ_{22}	0.45619	θ_{32}	0.90614
	(0.010431)		(0.11489)		(0.03899)
θ_{13}	0.025417	θ_{23}	-2.1391	θ_{33}	-0.071375
	(0.015212)		(0.8670)		(0.042555)
		θ_{24}	-0.89350		
			(0.70214)		

Table 3.4: Seemingly Unrelated Regression Estimates for Structural Model

State Space Estimates of Model (Equations 3.37 & 3.38)					
Parameter	Value	Parameter	Value	Parameter	Value
θ_{10}	0.1303	θ_{20}	-10.5889	θ_{30}	-0.7318
θ_{11}	0.9911	θ_{21}	2.7174	θ_{31}	0.1250
θ_{12}	-0.0056	θ_{22}	0.4893	θ_{32}	0.9077
θ_{13}	0.0313	θ_{23}	-1.9922	θ_{33}	-0.0689
		θ_{24}	-0.8033		

Table 3.5: State Space Estimates for Model

lagged GNE is extremely strong; and, the equation is effected by autocorrelation. Also, as would be expected, the interest rate equation results as presented in Table 3.2, are worse. The equation does not fit the data well, and the money supply parameter is not significant. The prices equation, as can be seen from Table 3.3, performs well, but, again, the money supply parameter is not significant.

But, the data is non-stationary[2], so that the estimates of standard errors may be biased implying that the tests of significance may not be valid. Being non-stationary also means that obtaining a good set of parameters for the model is more likely. This can be seen by examining the parameter estimates from the three estimation techniques. Comparing the Ordinary Least Squares estimates with those obtained by Seemingly Unrelated Regression (Table 3.4) and those generated by the State Space technique (Table 3.5) show that the parameter estimates are very similar. Table 3.4 includes an estimate of the (large sample) standard error for each parameter using the Seemingly Unrelated Regression technique. These are in parentheses under the estimated parameter. Again, they may be biased due to the non-stationary data. For this reason, no standard errors were produced with the State Space estimation technique.

The non-stationarity can also be seen via the econometric issue known as the 'unit root' problem. Note that the estimates of θ_{11} is close to one. Now if θ_{11} is one, then Equation 3.34

[2] Stationarity is here defined as the mean and variance of both the variables and the noise are constant with time.

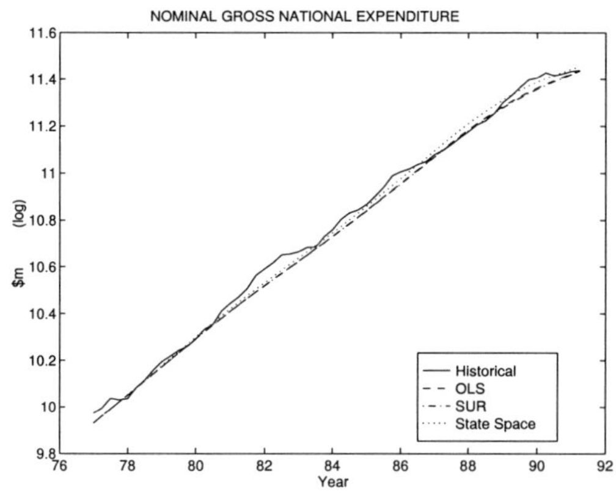

Figure 3.1: Comparison of Estimation Techniques for Gross National Expenditure

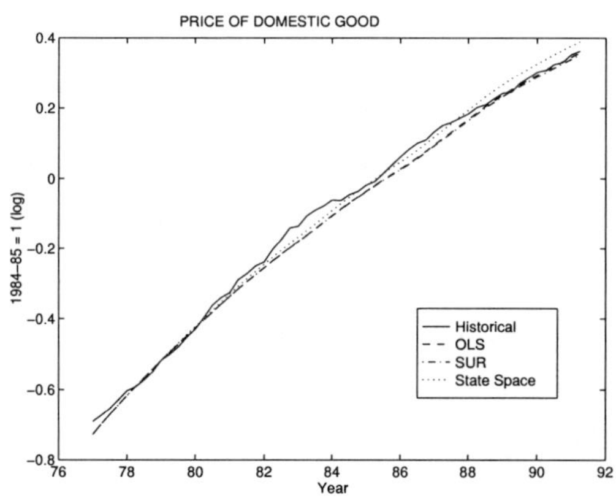

Figure 3.2: Comparison of Estimation Techniques for Prices

can be written as

$$Y_{t+1} = \theta_{10} + Y_t + W_t + v_{1,t} \tag{3.39}$$

with

$$W_t = \theta_{12} r_t + \theta_{13} F_t \tag{3.40}$$

Then by continual substitution back to initial conditions

$$Y_t = t\theta_{10} + Y_0 + \sum_{i=0}^{t-1} (W_i + v_{1,i}) \tag{3.41}$$

So, with $\mathcal{E}(.)$, being the mathematical expectation,

$$\mathcal{E}(Y_t) = t\theta_{10} + Y_0 + \sum_{i=0}^{t-1} W_i \tag{3.42}$$

and, so the variance

$$\mathcal{E}[Y_t - \mathcal{E}(Y_t)]^2 = \sum_{i=0}^{t-1} \mathcal{E}(v_{1,i}^2) = t\sigma_{v_1}^2 \tag{3.43}$$

Thus the variance of Y_t increases with t. The standard errors and t-values of the estimated coefficients will then be of little use.

Models with such problems are limited in their econometric usage if tests of significance are considered important. They are also of limited use in stochastic control exercises when good estimates are required for standard errors of the parameter values. In this book, the aim is to use models partly as low order observers for high order econometric models. The emphasis is on models that respond, for the inputs and outputs of interest, in a similar manner to the larger model. The low order model forms the basis of a controller. This may mean that the model may not need desirable econometric properties. Further such properties may even be lost from the model in a trade-off for improved control properties. This is a major advantage in using separate model based control.

Figure 3.1 visually compares the differing parameter techniques for national income. The Figure plots the historical data and the output of the model when run out from initial

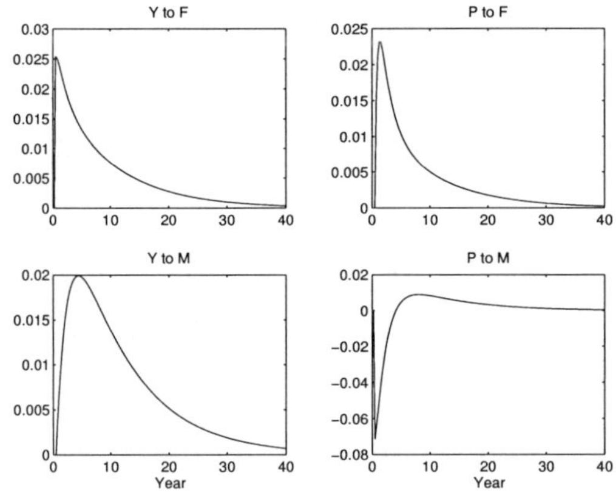

Figure 3.3: Model Multipliers

conditions and using historical values for the controls for each of the different parameter estimation techniques. Note that the models are run out from initial conditions so that previous model estimates are always used rather than the common econometric method of one step ahead prediction where previous values use historical data. Because of limited data, the prediction from initial conditions is for the same time horizon as was used to estimate the model. This is normal econometric practice. It can be seen, from the Figure, that firstly the model tracks the historical data well, and that results from each set of parameters are close to the historical data. Figure 3.2 is the analogous plots for prices. Again the model tracks the historical data well, and the models produced from the three parameter sets produce very similar outputs.

The parameters produce a model that is slow and stable. The model's behaviour can be seen by considering the impulse multipliers (or unit impulse response when initially at the origin). The multipliers generated the parameter set estimated by Seemingly Unrelated Regression are plotted in Figure 3.3. The Figure shows the multipliers are stable, small, slow and have no long run effects. A one Quarter unit impulse in fiscal policy increase both national income and prices, and this effect gradually decays over 20 years. Monetary policy, initially decreases prices and increases national income. Notice that the model supports the

OLS Estimates for Equation 3.44 (GNE)			
Parameter Name	Estimated Value	Standard Error	t-Value (51 d.f)
θ_{10}	0.10052	0.0519	2.0026
θ_{11}	-0.1338	0.1340	-0.99866
θ_{12}	1.1275	0.1348	8.3650
θ_{13}	-0.0062	0.0109	-0.5657
θ_{14}	0.0267	0.0159	1.6699
$R^2 = 0.9991$		$DW = 2.0396$	

Table 3.6: Six State Model: OLS Estimates for National Income

conventional wisdom that monetary policy is a more rapid control for prices and fiscal policy for national income.

The stability and slow response of the model can be examined by considering the eigenvalues of the state equation driving matrix. With the same parameter set, the eigenvalues are $\{0.98, 0.50, 0.88\}$ which are all in the unit circle but not close to the origin. Note the near unit root problem with the eigenvalue close to the unit circle. The model in state form is controllable and observable. In the closed loop the unit root will not be a problem, given the aim of generating a stable system.

3.3.6 *Six State Model*

Using the central difference approximation for the derivative generates a six state model with an extra lag for each of the dependent variables. Again, for estimation the model can be written in a structural form as[3]

$$Y_{t+1} = \theta_{10} + \theta_{11}Y_{t-1} + \theta_{12}Y_t + \theta_{13}r_t + \theta_{14}G_t + v_{1,t} \tag{3.44}$$

$$r_{t+1} = \theta_{20} + \theta_{21}r_{t-1} + \theta_{22}Y_t + \theta_{23}r_t + \theta_{24}M_t + \theta_{24}P_t + v_{2,t} \tag{3.45}$$

$$P_{t+1} = \theta_{30} + \theta_{31}P_{t-1} + \theta_{32}Y_t + \theta_{33}P_t + \theta_{34}M_t + v_{3,t} \tag{3.46}$$

[3] Parameters are, again, indicated by θ but they differ between the models.

OLS Estimates for Equation 3.45 (Interest Rates)			
Parameter Name	Estimated Value	Standard Error	t-Value (50 d.f)
θ_{20}	-11.566	6.4502	-1.7931
θ_{21}	-0.2177	0.1416	-1.5373
θ_{22}	3.0720	1.2204	2.5172
θ_{23}	0.5895	0.1381	4.2681
θ_{24}	-2.2740	0.9602	-2.3682
θ_{25}	-0.8498	0.7502	-1.1327
$R^2 = 0.6492$		$DW = 1.9617$	

Table 3.7: Six State Model: OLS Estimates for Interest Rates

In its state space form this model has the same controls and outputs, but the states are $\mathbf{x}_t = [Y_{t-1}, r_{t-1}, P_{t-1}, Y_t, r_t, P_t]^T$. The system matrices are

$$A = \begin{bmatrix} 0 & 0 & 0 & 1 & 0 & 0 \\ 0 & 0 & 0 & 0 & 1 & 0 \\ 0 & 0 & 0 & 0 & 0 & 1 \\ \theta_{11} & 0 & 0 & \theta_{12} & \theta_{13} & 0 \\ 0 & \theta_{21} & 0 & \theta_{22} & \theta_{23} & \theta_{25} \\ 0 & 0 & \theta_{31} & \theta_{32} & 0 & \theta_{33} \end{bmatrix} \quad B = \begin{bmatrix} 0 & 0 \\ 0 & 0 \\ 0 & 0 \\ \theta_{14} & 0 \\ 0 & \theta_{24} \\ 0 & \theta_{34} \end{bmatrix} \quad E = \begin{bmatrix} 0 \\ 0 \\ 0 \\ \theta_{10} \\ \theta_{20} \\ \theta_{30} \end{bmatrix} \quad H = \begin{bmatrix} 0 & 0 & 0 \\ 0 & 0 & 0 \\ 0 & 0 & 0 \\ 1 & 0 & 0 \\ 0 & 1 & 0 \\ 0 & 0 & 1 \end{bmatrix}$$

$$C = \begin{bmatrix} 0 & 0 & 0 & 1 & 0 & 0 \\ 0 & 0 & 0 & 0 & 0 & 1 \end{bmatrix} \qquad D = \begin{bmatrix} 0 & 0 \\ 0 & 0 \end{bmatrix}$$

The ordinary least squares parameter estimates for the national income structural equation (3.44) are given in Table 3.6; the interest rate equation (3.45) in Table 3.7; and the prices equation (3.46) in Table 3.8. As with the lower order model, many of the key theoretical features hold. Interestingly, none of the extra lags are statistically significantly different from zero at the usual confidence level. The other statistical properties are similar to the lower order model. But, due to the non-stationary data, these statistics may be biased.

The parameter estimates using the Three Stage Least Squares and the state space techniques

OLS Estimates for Equation 3.46 (Prices)			
Parameter Name	Estimated Value	Standard Error	t-Value (51 d.f)
θ_{30}	-0.64276	0.3487	-1.8433
θ_{31}	0.16751	0.1323	1.2665
θ_{32}	0.12168	0.0590	2.0608
θ_{33}	0.75181	0.1369	5.4924
θ_{34}	-0.07449	0.0442	-1.6860
$R^2 = 0.9994$		$DW = 1.8769$	

Table 3.8: Six State Model: OLS Estimates for Prices

3SLS Estimates of Parameters and Standard Errors for the Structural Model (Equations 3.44, 3.45 & 3.46)					
Parameter	Value	Parameter	Value	Parameter	Value
θ_{10}	0.099356 (0.047887)	θ_{20}	-13.334 (6.067)	θ_{30}	-0.64557 (0.33271)
θ_{11}	-0.13640 (0.12729)	θ_{21}	-0.17272 (0.12679)	θ_{31}	0.24169 (0.12078)
θ_{12}	1.1303 (0.1280)	θ_{22}	3.4703 (1.1372)	θ_{32}	0.12591 (0.05634)
θ_{13}	-0.0063665 (0.010379)	θ_{23}	0.51861 (0.12459)	θ_{33}	0.67742 (0.12552)
θ_{14}	0.025621 (0.015212)	θ_{24}	-2.5552 (0.8904)	θ_{33}	-0.079194 (0.042133)
		θ_{25}	-1.0033 (0.70654)		

Table 3.9: Six State Model: 3SLS Estimates for Structural Equations

State Space Estimates of Six State Model (Equations 3.37 & 3.38)					
Parameter	Value	Parameter	Value	Parameter	Value
θ_{10}	0.1072	θ_{20}	-12.0417	θ_{30}	-0.6389
θ_{11}	-0.1379	θ_{21}	-0.2289	θ_{31}	0.1642
θ_{12}	1.1324	θ_{22}	3.2334	θ_{32}	0.1200
θ_{13}	-0.0052	θ_{23}	0.5962	θ_{33}	-0.7554
θ_{14}	0.0306	θ_{24}	-2.4255	θ_{33}	-0.0729
		θ_{25}	-0.8686		

Table 3.10: Six State Model: State Space Estimates

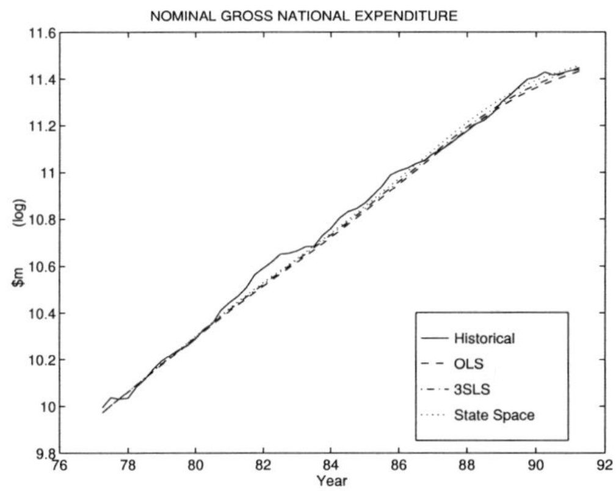

Figure 3.4: Six State Model: Comparison for Gross National Expenditure

are given in Tables 3.9 and 3.10 respectively. The parameter values are similar to each other, to the ordinary least squares parameters, and similar to the comparable parameters in the lower order model.

A comparison of the outputs from the various estimation techniques for the model is given in Figure 3.4 for the output of national income and Figure 3.5 for prices. The models generated by the different parameters sets produce good estimates of the historical data when run out from initial conditions and using historical values of the inputs. The national income estimates are better than the prices estimates with the relative error (using the 2-

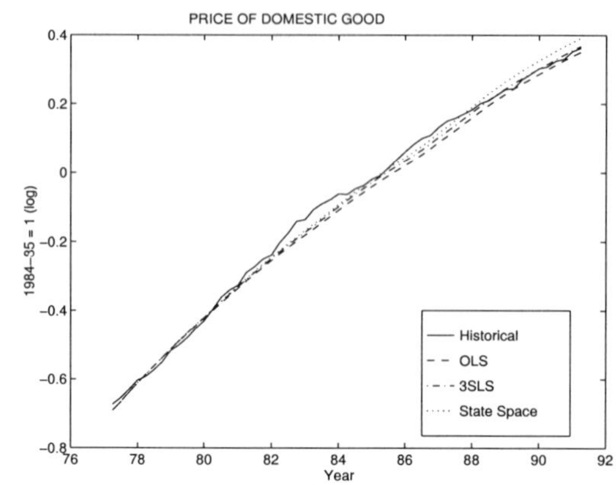

Figure 3.5: Six State Model: Comparison for Prices

norm) between the historical GNE and that generated by the model being 0.0026, 0.0028 and 0.0034 for the state space, Three Stage Least Squares and ordinary least squares techniques respectively. For prices the relative errors are 0.0636, 0.632 and 0.0815 respectively. The relative errors also show that it is better to estimate the model as a system rather than as individual equations.

This six state model is also slow and stable, as can be seen by the impact multipliers plotted in Figure 3.6. These use the parameters from the Three Stage Least Squares technique. In the state form, the model has eigenvalues of the state equation driving matrix of $\{0.2790 \pm 0.3123\imath, 0.1418, 0.9055, 0.9793, -0.2582\}$, so that it has a slight oscillatory response. The model is controllable and observable.

One of the adavntages of the six state model, especially when in state space form, is that it leads to an easy approximation for inflation. The inflation rate is often of more interest than the level of prices.

Now the inflation rate may be approximated by a first order Taylors series (backward

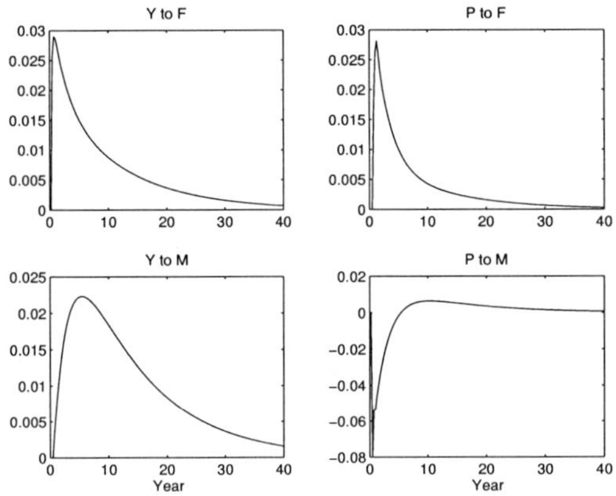

Figure 3.6: Six State Model: Multipliers

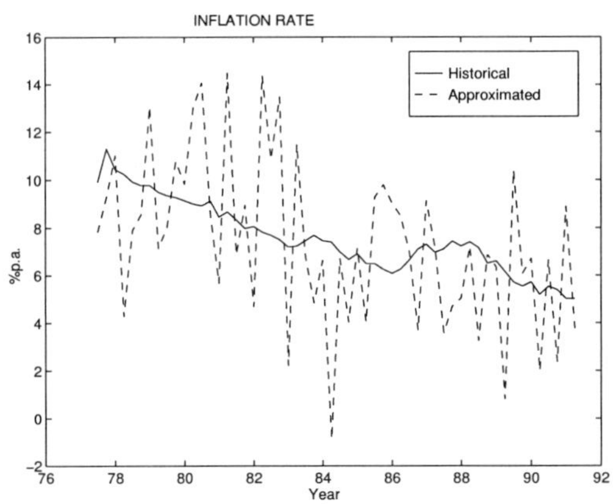

Figure 3.7: Six State Model: Approximating Inflation

differences) as

$$\frac{d \log P}{dt} \approx \frac{\log P_t - \log P_{t-1}}{\Delta t} \tag{3.47}$$

where, in this case, the time step, Δt, is a Quarter. Annualising this to a percentage inflation rate, gives

$$\frac{d \log P}{dt} \approx 400(\log P_t - \log P_{t-1}) \tag{3.48}$$

In the state space form of the model, both $\log P_t$ and $\log P_{t-1}$ are states, so the approximated inflation rate can be obtained by redefining the state output matrix to be

$$C = \begin{bmatrix} 0 & 0 & 0 & 1 & 0 & 0 \\ 0 & 0 & -400 & 0 & 0 & 400 \end{bmatrix} \tag{3.49}$$

The historical inflation rate and that generated by the model when run out from initial conditions and using historical controls are plotted in Figure 3.7. It can be seen that, while the model is reasonable for the overall trend, it does not predict the inflation rate well. This is not surprising as Equation 3.48 is in effect a smoothing operation. The relative error between the predicted and historical inflation rate is 0.3915.

3.4 Contemporaneous Control Time Series Models

The previous Sections presented models developed from theoretical structural frameworks. In this Section, models are built based on considering economic data as time series.

This Section specifically considers the issue of contemporaneous control in economic models and state space forms. Economic models are not usually expressed in the state space form used in this book, for, normally, no output equation is used. In this Section it is shown that an output equation can be used to overcome the problem of transforming contemporaneous control models into a state space form.

A second order, scalar difference equation with contemporaneous control is used as the starting point. This is expanded to a system of equations, to include longer lags, and, eventually, to include both exogenous variables and noise. It is shown that a number of state space forms can be generated for such models. Finally, using the approach, a low order econometric model is built.

In the analysis here z-transforms (with z being a complex variable) are used rather than shift operators as z-transforms have the advantages of allowing a full range of algebraic manipulation and allow for a direct relationship into continuous time systems. The latter is particularly relevant when the difference equations are generated from sampling of a continuous time system. Further, without loss of generality, the approach uses, pulse transfer functions (impulse multipliers) so that other effects (such as initial conditions) are zero.

3.4.1 Single Input Single Output Difference Equations

Consider the scalar, second order, contemporaneous control difference equation of

$$y_t = \alpha_1 y_{t-1} + \alpha_2 y_{t-2} + \beta_0 u_t \tag{3.50}$$

with initial conditions $y_{t-1} = \gamma_1, y_{t-2} = \gamma_2$. This is a single input, single output system.

Taking z-transforms for a pulse transfer function

$$\begin{aligned} Y(z) &= \alpha_1 z^{-1} Y(z) + \alpha_2 z^{-2} Y(z) + \beta_0 U(z) \\ &= \frac{\beta_0 U(z)}{1 - \alpha_1 z^{-1} - \alpha_2 z^{-2}} \end{aligned} \tag{3.51}$$

or, equivalently, by shifting forward in time

$$z^2 Y(z) = \alpha_1 z Y(z) + \alpha_2 Y(z) + \beta_0 z^2 U(z) \tag{3.52}$$

$$Y(z) = \frac{\beta_0 z^2 U(z)}{z^2 - \alpha_1 z - \alpha_2} \tag{3.53}$$

Using the transfer function of Equation 3.53, define $Z(z)$ such that

$$Z(z) = \frac{\beta_0 U(z)}{z^2 - \alpha_1 z - \alpha_2} \tag{3.54}$$

Then

$$Y(z) = z^2 Z(z) \tag{3.55}$$

Let

$$X_1(z) = Z(z) \tag{3.56}$$

$$X_2(z) = zZ(z) = zX_1(z) \tag{3.57}$$

Then

$$\begin{aligned} zX_2(z) &= z^2Z(z) \\ &= \alpha_1 zZ(z) + \alpha_2 Z(z) + \beta_0 U(z) \\ &= \alpha_1 X_2(z) + \alpha_2 X_1(z) + \beta_0 U(z) \end{aligned} \tag{3.58}$$

Thus

$$\begin{bmatrix} zX_1(z) \\ zX_2(z) \end{bmatrix} = \begin{bmatrix} 0 & 1 \\ \alpha_2 & \alpha_1 \end{bmatrix} \begin{bmatrix} X_1(z) \\ X_2(z) \end{bmatrix} + \begin{bmatrix} 0 \\ \beta_0 \end{bmatrix} U(z) \tag{3.59}$$

Now from Equations 3.55 and 3.58

$$Y(z) = \begin{bmatrix} \alpha_2 & \alpha_1 \end{bmatrix} \begin{bmatrix} X_1(z) \\ X_2(z) \end{bmatrix} + \beta_0 U(z) \tag{3.60}$$

Equations 3.59 and 3.60 give the model in state space form. Transforming back to the time domain, produces the discrete state space model of

$$\mathbf{x}_{t+1} = A\mathbf{x}_t + Bu_t \tag{3.61}$$

$$y_t = C\mathbf{x}_t + Du_t \tag{3.62}$$

with

$$A = \begin{bmatrix} 0 & 1 \\ \alpha_2 & \alpha_1 \end{bmatrix} \qquad B = \begin{bmatrix} 0 \\ \beta_0 \end{bmatrix}$$

$$C = \begin{bmatrix} \alpha_2 & \alpha_1 \end{bmatrix} \qquad D = \beta_0$$

and, after reversing the forward shift in time, with states and initial conditions of

$$\mathbf{x}_t = \begin{bmatrix} y_{t-1} \\ y_{t-2} \end{bmatrix} \qquad \mathbf{x}_0 = \begin{bmatrix} \gamma_1 \\ \gamma_2 \end{bmatrix} \tag{3.63}$$

The strength of this approach is that it easily scales up to systems of difference equations. But before doing so, this scalar example will be used to illustrate that this state form is not unique, and that a number of useful forms can be generated (Aoki, 1990; Åström and Wittenmark, 1990; Ogata, 1995).

Let P be any nonsingular matrix, and define the new state vector $\tilde{\mathbf{x}}_t = P\mathbf{x}_t$, then

$$\begin{aligned}
\tilde{\mathbf{x}}_{t+1} &= P\mathbf{x}_{t+1} \\
&= PA\mathbf{x}_t + PBu_t \\
&= PAP^{-1}\tilde{\mathbf{x}}_t + PBu_t \\
&= \tilde{A}\tilde{\mathbf{x}}_t + \tilde{B}u_t
\end{aligned} \tag{3.64}$$

and

$$\begin{aligned}
y_t &= C\mathbf{x}_t + Du_t \\
&= CP^{-1}\tilde{\mathbf{x}}_t + Du_t \\
&= \tilde{C}\tilde{\mathbf{x}}_t + Du_t
\end{aligned} \tag{3.65}$$

Importantly, this transformation has no effect on the eigenvalues of the state equation driving matrix as

$$\begin{aligned}
\mid \lambda I - A \mid &= \mid P^{-1} \mid\mid \lambda I - A \mid\mid P \mid \\
&= \mid \lambda I - P^{-1}AP \mid \\
&= \mid \lambda I - \tilde{A} \mid
\end{aligned} \tag{3.66}$$

with I being the 2×2 identity matrix. Thus the system can be transformed into a convenient form and the dynamics will be the same. Two common forms are the *controllable canonical form*, and the *observable canonical form*. The controllable canonical form is generated by not including β_0 in the definition of $Z(z)$ in Equation 3.54. In this case the states are the same as in the form above (Equation 3.63).

In controllable canonical form the system matrices are

$$\begin{aligned}
\breve{A} &= \begin{bmatrix} 0 & 1 \\ \alpha_2 & \alpha_1 \end{bmatrix} & \breve{B} &= \begin{bmatrix} 0 \\ 1 \end{bmatrix} \\
\breve{C} &= [\,\beta_0\alpha_2 \quad \beta_0\alpha_1\,] & D &= \beta_0
\end{aligned} \tag{3.67}$$

The observable canonical form has system matrices

$$\hat{A} = \begin{bmatrix} 0 & \alpha_2 \\ 1 & \alpha_1 \end{bmatrix} \qquad \hat{B} = \begin{bmatrix} \beta_0\alpha_2 \\ \beta_0\alpha_1 \end{bmatrix}$$
$$\hat{C} = [\,0 \quad 1\,] \qquad D = \beta_0 \tag{3.68}$$

The observable canonical form is mapped from the controllable canonical form by $\check{A}^T \to \hat{A}$, $\check{B}^T \to \hat{C}$ and $\check{C}^T \to \hat{B}$.

If the state equation driving matrix has distinct eigenvalues then the system can be transformed into *diagonal canonical form*, completely decoupling the equations. If the eigenvalues are not distinct, then the system can be transformed into a *Jordan canonical form*.

3.4.2 Multiple Input Multiple Output Difference Equations

The advantage of the approach presented here is that it easily scales up to vector difference equations. This is the case as long as the states are specified as variables ordered contiguously within time periods, rather than being, for example, ordered by variable.

Considering the multiple input multiple output, contemporaneous control, system of difference equations given by

$$\mathbf{y}_t = A_1\mathbf{y}_{t-1} + A_2\mathbf{y}_{t-2} + A_3\mathbf{y}_{t-3} + \cdots + A_q\mathbf{y}_{t-q} + B_0\mathbf{u}_t \tag{3.69}$$

with outputs $\mathbf{y} \in \Re^p$; controls (inputs) $\mathbf{u} \in \Re^m$; and coefficient matrices $B_0 \in \Re^{p\times m}$, and $A_i \in \Re^{p\times p}$ for $i = 1, 2, 3, \cdots, q$.

For this model (3.69), the pulse transfer function is given by

$$\mathbf{Y}(z) = [z^qI - z^{q-1}A_1 - z^{q-2}A_2 - z^{q-3}A_3 - \cdots - zA_{q-1} - A_q]^{-1}z^qB_0\mathbf{U}(z) \tag{3.70}$$

with I being the $p \times p$ identity matrix.

Define $\mathbf{Z}(z)$ such that

$$\mathbf{Z}(z) = [z^qI - z^{q-1}A_1 - z^{q-2}A_2 - z^{q-3}A_3 - \cdots - zA_{q-1} - A_q]^{-1}B_0\mathbf{U}(z) \tag{3.71}$$

Define states based on $\mathbf{Z}(z)$ but as contiguous blocks of the same degree as z, so that the output vector $\mathbf{y}$ is stacked up in time order. That is, let

$$\begin{aligned}
\mathbf{X_1}(z) &= \mathbf{Z}(z) \\
\mathbf{X_2}(z) &= z\mathbf{X_1}(z) = z\mathbf{Z}(z) \\
\mathbf{X_3}(z) &= z\mathbf{X_2}(z) = z^2\mathbf{Z}(z) \\
\mathbf{X_4}(z) &= z\mathbf{X_3}(z) = z^3\mathbf{Z}(z) \\
&\vdots \\
\mathbf{X_q}(z) &= z\mathbf{X_{q-1}}(z) = z^{q-1}\mathbf{Z}(z)
\end{aligned} \tag{3.72}$$

where $\mathbf{X_i} \in \Re^p, i = 1, 2, 3, \cdots, q$ are each ordered in the same sequence as $\mathbf{y}_t$.

The state equation can then be formed as

$$\begin{bmatrix} z\mathbf{X_1}(z) \\ z\mathbf{X_2}(z) \\ z\mathbf{X_3}(z) \\ z\mathbf{X_4}(z) \\ \vdots \\ z\mathbf{X_{q-1}}(z) \\ z\mathbf{X_q}(z) \end{bmatrix} = \begin{bmatrix} O_{p\times p} & I_{p\times p} & O_{p\times p} & O_{p\times p} & \cdots & O_{p\times p} & O_{p\times p} \\ O_{p\times p} & O_{p\times p} & I_{p\times p} & O_{p\times p} & \cdots & O_{p\times p} & O_{p\times p} \\ O_{p\times p} & O_{p\times p} & O_{p\times p} & I_{p\times p} & \cdots & O_{p\times p} & O_{p\times p} \\ \vdots & \vdots & \vdots & \vdots & \cdots & \vdots & \vdots \\ O_{p\times p} & O_{p\times p} & O_{p\times p} & O_{p\times p} & \cdots & O_{p\times p} & I_{p\times p} \\ A_q & A_{q-1} & A_{q-2} & A_{q-3} & \cdots & A_2 & A_1 \end{bmatrix} \begin{bmatrix} \mathbf{X_1}(z) \\ \mathbf{X_2}(z) \\ \mathbf{X_3}(z) \\ \mathbf{X_4}(z) \\ \vdots \\ \mathbf{X_{q-1}}(z) \\ \mathbf{X_q}(z) \end{bmatrix} + \begin{bmatrix} O_{p\times m} \\ O_{p\times m} \\ O_{p\times m} \\ O_{p\times m} \\ \vdots \\ O_{p\times m} \\ B_0 \end{bmatrix} \mathbf{U}(z) \tag{3.73}$$

with $I_{p\times p}$ being the $p \times p$ identity matrix and $O_{p\times p}$ being the $p \times p$ zero matrix.

As $\mathbf{Y}(z) = z^q\mathbf{Z}(z)$, the output equation is

$$\mathbf{Y}(z) = [\, A_q \quad A_{q-1} \quad A_{q-2} \quad A_{q-3} \quad \cdots \quad A_2 \quad A_1 \,] \begin{bmatrix} \mathbf{X_1}(z) \\ \mathbf{X_2}(z) \\ \mathbf{X_3}(z) \\ \mathbf{X_4}(z) \\ \vdots \\ \mathbf{X_{q-1}}(z) \\ \mathbf{X_q}(z) \end{bmatrix} + B_0\mathbf{U}(z) \tag{3.74}$$

The state and output equations can be written in the standard state space form as

$$z\mathbf{X}(z) = A\mathbf{X}(z) + B\mathbf{U}(z) \tag{3.75}$$

$$\mathbf{Y}(z) = C\mathbf{X}(z) + D\mathbf{U}(z) \tag{3.76}$$

where $\mathbf{X} \in \Re^n$; $A \in \Re^{n\times n}$; $B \in \Re^{n\times m}$; $C \in \Re^{p\times n}$; $D \in \Re^{p\times m}$; and, $n = mq$.

The multipliers for the system can then be written as

$$\mathbf{Y}(z) = (C(zI_{n\times n} - A)^{-1}B + D)\mathbf{U}(z) \tag{3.77}$$

This generates the same multipliers as Equation 3.70.

In the time domain, the state space model is then

$$\mathbf{x}_{t+1} = A\mathbf{x}_t + B\mathbf{u}_t \tag{3.78}$$

$$\mathbf{y}_t = C\mathbf{x}_t + D\mathbf{u}_t \tag{3.79}$$

with $\mathbf{x}_t = [\mathbf{y}_{t-1}, \mathbf{y}_{t-2}, \mathbf{y}_{t-3}, \mathbf{y}_{t-4}, \cdots, \mathbf{y}_{t-q+1}, \mathbf{y}_{t-q}]^T$.

3.4.3 VAR Models

It is common for multiple input, multiple output, recursive economic models to also include level constants and noise. Such vector autoregressive (VAR) models with contemporaneous control can be written as

$$\mathbf{y}_t = \mathbf{e} + A_1\mathbf{y}_{t-1} + A_2\mathbf{y}_{t-2} + A_3\mathbf{y}_{t-3} + \cdots + A_q\mathbf{y}_{t-q} + B_0\mathbf{u}_t + \mathbf{v}_t \qquad (3.80)$$

where $\mathbf{e} \in \Re^p$ is a time invariant vector of level constants, and $\mathbf{v} \in \Re^p$ is a vector of noise (or error terms).

This type of model can also be written in the state space form by, again, creating states of the stacked up over time output variables ($\mathbf{y}$). This is used to generate the state equation. The output equation is, again, formed by the bottom partition slice (i.e. last p individual equations) of the state equation.

Let

$$E = \begin{bmatrix} O_{p\times 1} \\ O_{p\times 1} \\ O_{p\times 1} \\ O_{p\times 1} \\ \vdots \\ O_{p\times 1} \\ \mathbf{e} \end{bmatrix} \qquad H = \begin{bmatrix} O_{p\times p} \\ O_{p\times p} \\ O_{p\times p} \\ O_{p\times p} \\ \vdots \\ O_{p\times p} \\ I_{p\times p} \end{bmatrix} \qquad (3.81)$$

Then the state space form is given by

$$\mathbf{x}_{t+1} = A\mathbf{x}_t + B\mathbf{u}_t + E + H\mathbf{v}_t \qquad (3.82)$$

$$\mathbf{y}_t = C\mathbf{x}_t + D\mathbf{u}_t + \mathbf{e} + \mathbf{v}_t \qquad (3.83)$$

The following is a numerical example of the two period lagged VAR model of the type given in Equation 3.80. The model is a two input, two output model, with parameters and initial conditions given by

$$A_1 = \begin{bmatrix} 0.1 & 0.2 \\ 0.3 & 0.2 \end{bmatrix} \quad A_2 = \begin{bmatrix} 0.3 & 0.1 \\ 0.1 & 0.2 \end{bmatrix} \quad B_0 = \begin{bmatrix} 0.5 & 0.6 \\ 0.4 & 0.2 \end{bmatrix} \quad \mathbf{e} = \begin{bmatrix} 1 \\ 2 \end{bmatrix}$$
$$\mathbf{y}_2 = \begin{bmatrix} 0.2 \\ 0.4 \end{bmatrix} \quad \mathbf{y}_1 = \begin{bmatrix} 0.1 \\ 0.3 \end{bmatrix} \qquad (3.84)$$

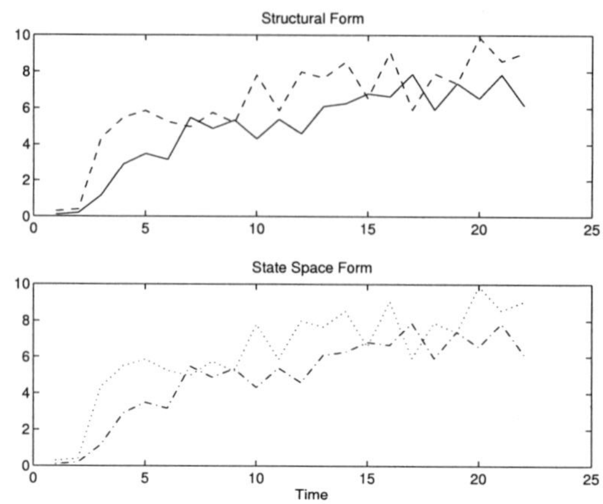

Figure 3.8: Output from Numerical Example

In state space form the system matrices are then

$$A = \begin{bmatrix} 0 & 0 & 1 & 0 \\ 0 & 0 & 0 & 1 \\ 0.3 & 0.1 & 0.1 & 0.2 \\ 0.1 & 0.2 & 0.3 & 0.2 \end{bmatrix} \quad B = \begin{bmatrix} 0 & 0 \\ 0 & 0 \\ 0.5 & 0.6 \\ 0.4 & 0.2 \end{bmatrix} \quad E = \begin{bmatrix} 0 \\ 0 \\ 1 \\ 2 \end{bmatrix} \quad H = \begin{bmatrix} 0 & 0 \\ 0 & 0 \\ 1 & 0 \\ 0 & 1 \end{bmatrix}$$

$$\mathbf{x}_3 = \begin{bmatrix} 0.1 \\ 0.3 \\ 0.2 \\ 0.4 \end{bmatrix} \quad C = \begin{bmatrix} 0.3 & 0.1 & 0.1 & 0.2 \\ 0.1 & 0.2 & 0.3 & 0.2 \end{bmatrix} \quad D = \begin{bmatrix} 0.5 & 0.6 \\ 0.4 & 0.2 \end{bmatrix} \tag{3.85}$$

Figure 3.8 plots the output from the model with input ($\mathbf{u}$) of a random sequence evenly distributed between zero and one; and, with the noise ($\mathbf{v}$) being independent and normally distributed each with unit variance. The upper portion of the Figure plots the outputs in structural form, and the lower portion in state space form. The outputs are identical.

Vector autoregressive models such as that given by Equation 3.80 are often referred to as VARX models, as the controls, in the open loop, are exogenous. Lütkepohl (1991, p.430), for example, transforms such models into a state space form but the exogenous variables are left as contemporaneous with the state update in the state equation. That state space cannot be used with standard control theory algorithms. However, the state space form used in this book has the control in the output equation, and so incorporates contemporaneous control within the standard control engineering state space form. This allows for the use of standard control theory algorithms.

A further problem with models of the form given by Equation 3.80, is that they may suffer from non-stationarity. In this case the statistical properties can be improved if the outputs ($\mathbf{y}$) are cointegrated (Johansen and Juselius, 1991), and then Equation 3.80 can be rewritten into the form

$$\begin{aligned}\Delta\mathbf{y}_t =& \acute{\mathbf{e}} + D_1\Delta\mathbf{y}_{t-1} + D_2\Delta\mathbf{y}_{t-2} + D_3\Delta\mathbf{y}_{t-3} + \cdots + D_{q-1}\Delta\mathbf{y}_{t-q+1} \\ & - \Pi\mathbf{y}_{t-q} + \acute{B}_0\Delta\mathbf{u}_t + \acute{\mathbf{v}}_t\end{aligned} \tag{3.86}$$

where

$$\Delta\mathbf{y}_t = \mathbf{y}_t - \mathbf{y}_{t-1} \tag{3.87}$$

$$D_i = -(I - A_1 - \cdots - A_i), \qquad i = 1, 2, \cdots, q-1 \tag{3.88}$$

$$\Pi = \alpha\beta^T = I - A_1 - A_2 - \cdots - A_q \tag{3.89}$$

with I being the $p \times p$ identity matrix, and $D_i, \Pi \in \Re^{p\times p}$.

The degree of cointegration is given by the rank of Π, say r, with $r < p$. Here $\alpha, \beta \in \Re^{p\times r}$; each column of β, the cointegrating matrix, gives the coefficients of a cointegrating relationship between the outputs; and each row of α, the loading matrix, gives the weights that are applied to the different cointegrating relationship for each output.

Equation 3.86 is best considered as deviations about the steady state output relationship given by the term $\Pi\mathbf{y}_{t-q}$, and is essentially the same as that frequently used in the control literature as deviations about some operating level.

Equation 3.86 can also be written in the same state space form of Equations 3.82 and 3.83. In this case, the states become the first differences of the model outputs ($\Delta\mathbf{y}$), the controls

OLS Estimates for Inflation			
Parameter Name	Estimated Value	Standard Error	t-Value (44 d.f)
$A_{1(1,1)}$	-0.083714	0.1569	-0.53354
$A_{1(1,2)}$	1.1760	1.4138	0.83178
$A_{2(1,1)}$	0.15464	0.1640	0.94277
$A_{2(1,2)}$	-2.7105	2.4870	-1.0899
$A_{3(1,1)}$	-0.043916	0.1475	-0.29773
$A_{3(1,2)}$	-0.51540	2.4991	-0.20623
$A_{4(1,1)}$	0.22738	0.1492	1.5238
$A_{4(1,2)}$	0.66088	1.6163	0.40899
$B_{0(1,1)}$	5.4415	4.7566	1.1440
$B_{0(1,2)}$	68.266	55.4991	1.2300
$R^2 = 0.3368$		$DW = 2.0388$	

Table 3.11: Stationary VAR Model: OLS Estimates for Inflation

are the first differences of the model inputs, and $\mathbf{e}$ is replaced by $\acute{\mathbf{e}} - \Pi\mathbf{y}_{t-p}$. That is, the same state space form can be used in this case.

The state space form of Equations 3.82 and 3.83 can be used for the levels form (Equation 3.80) and for the first differences form (Equation 3.86). Algorithms that use the state space form can then be used with either model.

3.4.4 Econometric Models

The approach can be used to build contemporaneous control time series based econometric models. Firstly, a stationary data, two input, two output, four period lagged model is built. This model is designed to be different from previous models developed in this Chapter. The outputs are the inflation rate, and the unemployment rate; and, the inputs are a monetary and a fiscal policy control.

The data series are the same as used in earlier models in this Chapter (see Section 3.3.4 on page 74). The outputs are the inflation rate (INF) measured as a percentage per annum,

OLS Estimates for Unemployment			
Parameter Name	Estimated Value	Standard Error	t-Value (44 d.f)
$A_{1(2,1)}$	0.014943	0.0176	0.84982
$A_{1(2,2)}$	1.3796	0.1584	8.7075
$A_{2(2,1)}$	-0.008046	0.0184	-0.43769
$A_{2(2,2)}$	-0.28705	0.2787	-1.0299
$A_{3(2,1)}$	0.001157	0.0165	0.06998
$A_{3(2,2)}$	-0.07795	0.2801	-0.27835
$A_{4(2,1)}$	-0.15388	0.0167	-0.91952
$A_{4(2,2)}$	-0.10763	0.1811	-0.59425
$B_{0(2,1)}$	0.08688	0.5330	0.16299
$B_{0(2,2)}$	-7.4023	6.2194	-1.1902
$R^2 = 0.9373$		$DW = 2.0581$	

Table 3.12: Stationary VAR Model: OLS Estimates for Unemployment

and the unemployment rate also measured as a percentage (URT). With the four period lag, a full year of observations is included in the model. The controls are, again, the Target Ratio of Public Debt to Gross Domestic Product (RPUBDE) for fiscal policy, and the Nominal Money Supply (M) for the monetary policy. For this model each series has been normalised to remove the need for level constants.

All the data series except the money supply are stationary. A common transform to make a data series stationary is to take first differences of logarithms, though it has been suggested (Lütkepohl, 1991, p.350) that taking differences may distort the data. This transformation turns level data series into percentage changes. The transformation is used here, so that the monetary policy control becomes percentage change in the Nominal Money Supply ($\Delta \log \mathrm{M}$). The percentage change in the money supply is frequently used as a monetary policy instrument (e.g. Murphy, 1996*b*).

In VAR form, the model can be written as

$$\mathbf{y}_t = A_1\mathbf{y}_{t-1} + A_2\mathbf{y}_{t-2} + A_3\mathbf{y}_{t-3} + A_4\mathbf{y}_{t-4} + B_0\mathbf{u}_t + \mathbf{v}_t \tag{3.90}$$

where $A_i, B_0 \in \Re^{2\times 2}, i = 1,2,3,4$; $\mathbf{y}_t = [\mathrm{INF}_t, \mathrm{URT}_t]^T$; $\mathbf{u}_t = [\mathrm{RPUBDE}_t, \Delta \log \mathrm{M}_t]^T$; and, $\mathbf{v}_t \in \Re^2$ are the error terms.

Again, the model was estimated using Ordinary Least Squares for each of the individual equations of Equation 3.90; jointly using Three Stage Least Squares; and in state space form. In the latter case, again, the Ordinary Least Squares parameters were used as the starting point for an optimisation procedure which aims to minimise the sum of the 2-norm of the difference between the historical data and that generated when the model is run out from initial conditions for each of the outputs.

The Ordinary Least Squares parameter estimates and their statistics are presented in Tables 3.11 and 3.12 for the inflation and unemployment equations respectively. Table 3.11 shows that the inflation equation poorly explains the variations in the data. More importantly, none of the explanatory variables are significant at the usual level of confidence. Perhaps structural models based on economic theory are preferable than time series models when considering inflation. The results of the unemployment equation are similar to the inflation equation. Here, though, there is a better fit of the data (as measured by the R^2 statistic). But only the one period lagged unemployment rate is significant.

The resultant parameters from the Three Stage Least Squares estimation are given in Tables 3.13 and 3.14. The Tables give the estimated parameters and estimate of the (large sample) standard errors. These results differ little from those obtained using Ordinary Least Squares.

The parameters obtained from the state space estimation procedure, in the form of Equation 3.90, are

$$\hat{A}_1 = \begin{bmatrix} -0.4223 & 0.4329 \\ 0.0348 & 1.9173 \end{bmatrix} \quad \hat{A}_2 = \begin{bmatrix} 0.7520 & -1.4807 \\ -0.0195 & -0.7514 \end{bmatrix} \quad \hat{A}_3 = \begin{bmatrix} 0.1457 & -1.5198 \\ 0.0057 & -0.6616 \end{bmatrix}$$

$$\check{A}_4 = \begin{bmatrix} 0.5536 & 2.4876 \\ 0.1357 & 0.5893 \end{bmatrix} \qquad \hat{B}_0 = \begin{bmatrix} -0.0820 & 72.4773 \\ -0.0409 & -7.9149 \end{bmatrix}$$

3SLS Estimates for Inflation		
Parameter Name	Estimated Value	Standard Error
$A_{1(1,1)}$	-0.083714	0.15690
$A_{1(1,2)}$	1.1760	1.4138
$A_{2(1,1)}$	0.15464	0.16403
$A_{2(1,2)}$	-2.7105	2.4870
$A_{3(1,1)}$	-0.043916	0.14750
$A_{3(1,2)}$	-0.51540	2.4991
$A_{4(1,1)}$	0.22738	0.14922
$A_{4(1,2)}$	0.66088	1.6163
$B_{0(1,1)}$	5.4415	4.7566
$B_{0(1,2)}$	68.266	55.499

Table 3.13: Stationary VAR Model: 3SLS Estimates for Inflation

3SLS Estimates for Unemployment		
Parameter Name	Estimated Value	Standard Error
$A_{1(2,1)}$	0.014943	0.017583
$A_{1(2,2)}$	1.3796	0.15843
$A_{2(2,1)}$	-0.0080456	0.018382
$A_{2(2,2)}$	-0.28705	0.27870
$A_{3(2,1)}$	0.0011567	0.016529
$A_{3(2,2)}$	-0.077954	0.28006
$A_{4(2,1)}$	-0.15376	0.016722
$A_{4(2,2)}$	-0.10763	0.18112
$B_{0(2,1)}$	0.086880	0.53303
$B_{0(2,2)}$	-7.4023	6.2194

Table 3.14: Stationary VAR Model: 3SLS Estimates for Unemployment

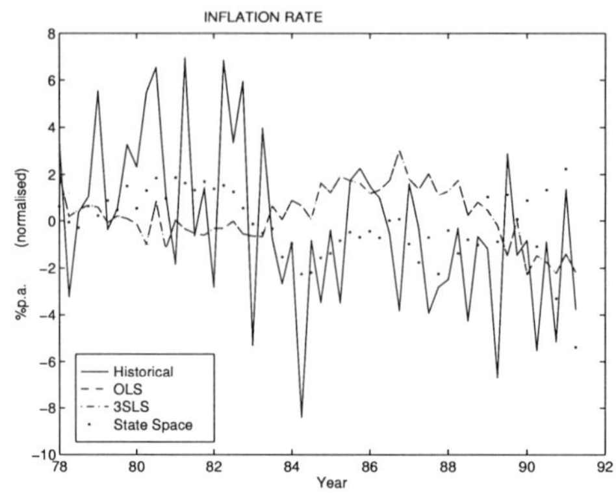

Figure 3.9: Output for Inflation Rate

These parameters differ from those obtained by the conventional econometric techniques.

In state space form, the model is a two input, two output, eight state model. With all parameter sets it is controllable and observable. The state equation driving matrix has eigenvalues of $\{0.8180 \pm 0.2368\imath, 0.8148, -0.8140, -0.0373 \pm 0.6314\imath, -0.1332 \pm 0.2379\imath\}$ for the Ordinary Least Squares and Three Stage Least Squares parameters; and $\{-1.2344, -0.2316 \pm 0.4684\imath, 0.9846 \pm 0.1644\imath, -0.3961, 0.8003, -0.1073\}$ for the state space parameters. Note that the model with the Ordinary Least Squares parameters is stable, but with the state space parameters is unstable and that this model has an eigenvalue close to one (i.e. a near 'unit root'). The state space parameters produce a less well conditioned state equation driving matrix ($\kappa_2(A_{SS}) = 1846$), than does the Ordinary Least Squares parameters ($\kappa_2(A_{LS}) = 325$).

The resultant output from the different parameter sets and the historical data are presented visually in Figure 3.9 and 3.10. In the Figures the model is run out in state space form from initial conditions, so the lagged variables are those predicted by the model rather than being historical data. The Figures show the better fit of the model when parameters are estimated using the state space procedure, especially for the unemployment rate (Figure 3.10). The relative error (using the 2-norm) between the historical data for the inflation rate and that

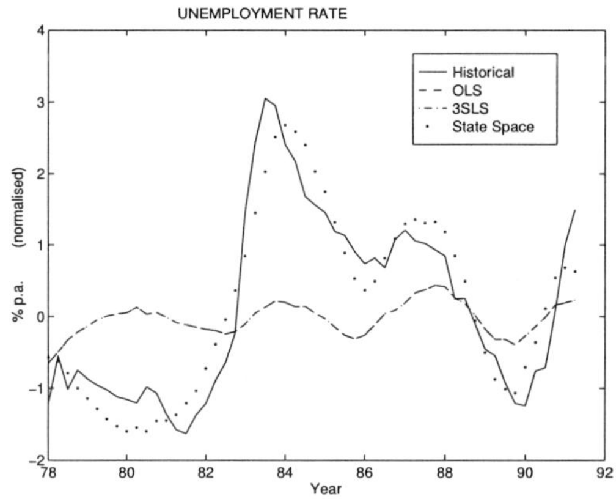

Figure 3.10: Output for Unemployment Rate

estimated with the parameter sets are 1.07 and 0.85 for Ordinary Least Squares and state space techinques respectively. For the unemployment rate, the difference is significant with the reduction from 0.93 for Ordinary Least Squares to 0.35 for the state space technique.

The final model has the same structure as the above model, so that it has two outputs each of which are lagged for a year, and has two contemporaneous inputs. But, in this case, the model uses the stock variables of Nominal Gross National Expenditure (GNEz) and Total Employment (NT) for outputs; and, for inputs, the Money Supply (M) and General Government Purchases of Consumption Goods (GGCO) as the monetary and fiscal policy controls, respectively. Logarithms of all data series are used. The series are normalised by the subtraction of their mean over the sample time horizon as this removes the need for intercepts with least squares estimation. The model can be written in the form of Equation 3.90.

The results of stationarity tests are given in Table 3.15. The Augmented Dickey-Fuller Tests are on the basis of no intercepts and no time trends. The Table shows that, at the 10% (2-tailed) confidence level, the hypothesis of non-staionarity cannot be rejected for the data series. If the data series are non-stationary but the residuals are stationary,

Augmented Dickey-Fuller Statistic	
Data Series	t-statistic
Gross National Expenditure	-1.8493
Total Employment	-0.7675
Government Consumption	-0.9503
Money Supply	-1.0327
Residuals	-3.2631
OLS GNE Residuals	-5.3487
OLS Employment Residuals	-5.1456
Critical Value	-4.42

Table 3.15: Stationarity Test Statistics

Estimation of the Model				
Parameter Name	Ordinary Least Squares	Three Stage Least Squares	State Space	Standard Error (OLS & 3SLS)
$A_{1(1,1)}$	0.95685	0.95685	0.7041	0.1478
$A_{1(1,2)}$	-0.16378	-0.16378	-0.0759	0.4451
$A_{1(2,1)}$	0.069543	0.069543	0.1036	0.05109
$A_{1(2,2)}$	1.1509	1.1509	0.7824	0.1538
$A_{2(1,1)}$	0.032328	0.032328	-0.0113	0.2030
$A_{2(1,2)}$	-0.22509	-0.22509	-1.3663	0.66210
$A_{2(2,1)}$	-0.13691	-0.13691	-0.0938	0.07016
$A_{2(2,2)}$	0.069961	0.069961	0.0032	0.2288
$A_{3(1,1)}$	-0.025675	-0.025675	-0.2688	0.2172
$A_{3(1,2)}$	-0.13159	-0.13159	1.6343	0.6135
$A_{3(2,1)}$	-0.048005	-0.048005	-0.2108	0.07505
$A_{3(2,2)}$	-0.21846	-0.21846	0.9012	0.2120
$A_{4(1,1)}$	-0.25303	-0.25303	0.0596	0.1296
$A_{4(1,2)}$	0.39587	0.39587	-0.6077	0.3984
$A_{4(2,1)}$	0.071133	0.071133	0.1420	0.04478
$A_{4(2,2)}$	-0.076067	-0.076067	-0.6957	0.1377
$B_{0(1,1)}$	0.013221	0.013221	-0.0256	0.01554
$B_{0(1,2)}$	0.29781	0.29781	0.6057	0.07001
$B_{0(2,1)}$	-0.0036134	-0.0036134	0.0171	0.005370
$B_{0(2,2)}$	0.061674	0.061674	0.0466	0.02422

Table 3.16: Non-stationary VAR Model: Parameter Estimates

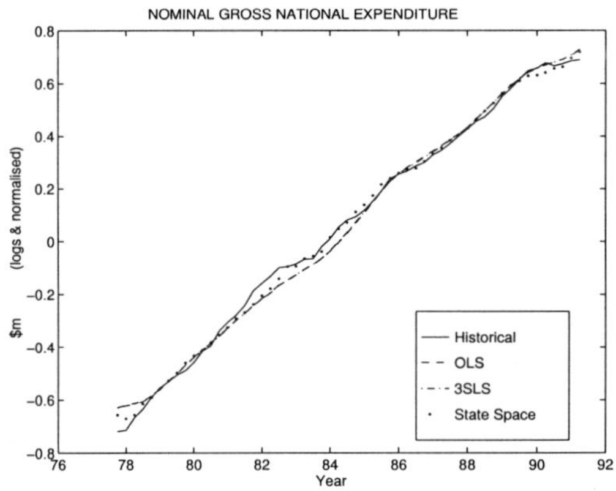

Figure 3.11: Output for Gross National Expenditure

then the series are cointegrated and the usual regression statistics may not be biased. The Augmented Dickey-Fuller Test on the residuals shows that the null hypothesis of no cointegrating relationship between the data series cannot be rejected, but as the test statistic is reasonably close to the critical value, it could be that a Type-2 error may be made in concluding there is no cointegration. Further, as the residuals from the Ordinary Least Squares estimated equations for the outputs cannot be considered non-stationary by the Augmented Dickey-Fuller Test, then faith can be held in their relationship over the long term. The model is, then, probably reasonable.

The parameter estimates from the three techniques are presented in Table 3.16. It can be seen from the Table that those from Ordinary Least Squares and those from Three Stage Least Squares are identical. Standard errors are also presented in the Table for the Ordinary Least Squares and Three Stage Least Squares techniques. Due to questions of stationarity care must be taken in interpreting these statistics. As the standard errors for the Ordinary Least Squares parameters are identical to those for the Three Stage Least Squares parameters, standard errors are presented in a separate column.

In state space form, the model is a two input, two output, eight state model. With all

Figure 3.12: Output for Employment

parameter sets it is controllable and observable. The state equation driving matrix has eigenvalues of $\{-0.4129 \pm 0.4903\imath, -0.3673, 0.0676, 0.7399 \pm 0.3534\imath, 0.9186 \pm 0.1575\imath\}$ for the Ordinary Least Squares parameters; and $\{-0.5250 \pm 0.6032\imath, -0.0777 \pm 0.5014\imath, 0.4650 \pm 0.3834\imath, 0.7190, 1.0431\}$ for the state space parameters. Again, the model with the Ordinary Least Squares parameters is stable, but with the state space parameters is unstable. The near unit eigenvalue ('unit root') indicating non-stationarity can also be seen. The state space parameters produce a better conditioned state equation driving matrix ($\kappa_2(A_{SS}) = 98$), than does the Ordinary Least Squares parameters ($\kappa_2(A_{LS}) = 222$), though neither matrix is well-conditioned.

The resultant output from the different parameter sets and the historical data are presented visually in Figure 3.11 and 3.12. In the Figures the model is run out in state space form from initial conditions. The Figures show the better fit of the parameters estimated using the state space procedure. The relative error (using the 2-norm) between the historical data for Gross National Expenditure and that estimated with the parameter sets are 0.078 and 0.051 for Ordinary Least Squares and state space techinques respectively. For employment, the values are 0.315 for Ordinary Least Squares to 0.125 for the state space technique.

3.5 Conclusion

The purpose of this Chapter has been to build a suite of economic models. The focus has been upon low order dynamic macroeconometric models. The Chapter has shown how a range of models, built using a variety of techniques, can be transformed into a standard state space form. This state space form can be used with the standard control algorithms. It differs from that usually used in the economics control literature in that it contains an output equation, and differs from that in the control engineering literature in that it contains level constants and exogenous variables.

The suite of models contains continuous time, discrete time, linear and nonlinear models; though the emphasis has been on discrete time, linear models. A specific issue with discrete time models is that of contemporaneous control. Such models are common in economics, and have created difficulties in transforming models into state space form (Chow, 1975; Kendrick, 1981). This Chapter has shown how these difficulties can be overcome.

The first model built in this Chapter was a Multiplier Accelerator model. This is a standard economic model that has been frequently used in economic control exercises. It is a contemporaneous control model.

The second class of models were developed from first principles using the IS-LM theoetical framework. From this nonlinear equilibrium framework, a linear, continuous time model was built. This was extended to a model to explain the dynamic adjustment between equilibrium positions. Extending the IS-LM framework through the inclusion of an expectations augmented Phillips curve, introduced prices into the model. Finally, the model was transformed into a number of discrete time forms.

The third class of models were based upon time series techniques rather than economic theory. These models are Vector Autoregressive models, and are contemporaneous control models.

The models developed in this Chapter have been estimated using Quarterly Australian data. Standard econometric estimation techniques, including Ordinary Least Squares and

Three Stage Least Squares have been used. Econometric issues, such as non-stationarity and cointegration, have also been considered. A further parameter estimation routine was used for estimating the models in state space form. This routine makes specific use of the output equation.

One of the difficulties of economic modelling is the problem with data. Economic data is difficult to obtain, so the practice of model verificiation by comparing the model predicitions using a different set of data from that used in parameter estimation, is not common. Further, model verification via statistical tests on parameters may be misleading using macroeconomic data, for that data is likely to be non-stationary.

The Chapter has developed a suite of models which are used in later Chapters.

Chapter 4

CONTROL STRATEGIES FOR ATTAINING RAMP OUTPUT

4.1 Introduction

This Chapter considers the issue of tracking a desired output of a set of ramps. A set of outputs is used as, for most economic systems, it is the trade-offs between multiple inputs and multiple outputs that are of interest. Ramps follow naturally with economic systems, as many economic data series have a ramp trajectory over time periods of interest, and, thus, economic models tend to be expressed so that ramps are a common output. As it is an aim of this book to use standard economic models, then ramps follow naturally. This is especially the situation when using low order observers for large nonlinear macroeconometric models.

Figures 3.1 and 3.2 (on page 79), plot the historical data for Australian Gross National Expenditure and the price level, respectively. These graphs clearly show the ramp trend of these economic indicators. Many other macroeconomic variables also follow such ramp trajectories. When using such data for the development of economic models, a common practice is to attempt to transform the data to be stationary. The usual transformation is first difference of logarithms. While this may improve the statistical properties of the model, it adds a complication if the model is used as an observer for a higher order nonlinear model. For this reason, the model output in levels is used here.

This Chapter focuses on the development of control strategies necessary to track desired outputs of a set of ramps. Single input single output systems of System Type 2 or above will track a ramp with no steady state error (Ogata, 1990; León De La Barra S. and Fernández F., 1994). Here the primary concern is with multiple input multiple output systems, and the aim is to track a vector of ramps. In this case System Type can be calculated between each input and output, but not for a transfer function for the whole system. Further, the aim is not to limit models to be of a particular System Type.

Three different control methods of ramp tracking are considered in this Chapter. In Section 4.3 a method based on the addition of integrators is used; and, in Section 4.4 a method based on feedforward control is developed. The third method, in Section 4.5 is the standard linear quadratic output tracking method. The methods are developed using continuous time and discrete time, and are illustrated by using low order economic models considered in earlier Chapters. Parameter uncertainty is a significant issue with economic systems, so the Chapter particularly examines the implications of parameter perturbation on the closed loop system.

The Integral method developed in this Chapter extends the control engineering servo tracking problem into a multiple input, multiple output, ramp tracking setting. Importantly, it shows how this can be done by extending the techniques that are in common usage in servo tracking problems. It also shows how the method can be extended to track desired outputs that are of higher order polynomial in time. The advantages and disadvantages of the method, and the other methods, are considered.

The Feedforward method developed in this Chapter also extends the standard control engineering method to multiple input, multiple output, ramp tracking situations. It shows how the standard computational technique of fixed point iteration can be used in the solution of the ramp case.

Both of these methods directly effect the structure of the model, and probably do not make economic sense. For this book, altering the economic properties of the model need not be a problem for the models are used as observers to economic systems. This is consistent with a major theme of this book which is that control policies are generated on separate objects from the economic system object that is subject to these policies. It is the policy analysis on the economic system object, such as a macroeconometric model, that is important, not the controller. The economic system object needs to make economic sense not necessarily the control model, for it is just another (but hopefully better) way of generating economic policy.

The Linear Quadratic Output Tracking method is a standard control engineering method. Its use in economics is limited, for output equations are not usually used. In economics, it is the linear quadratic state tracking problem that is normally considered. Frequently, though, some states are given zero weights in the objective function. Zero weights can be implemented more easily, and with less possibility of the introduction of numeric errors, through an output equation.

The state space form used in this Chapter allows for the use of economic models that contain contemporaneous controls.

Parameter uncertainty is a major issue with economic models, and the effects of this uncertainty on the formation of optimal control rules has been examined by Aoki (1967) and Kendrick (1981). In that case, specific assumptions about the statistical properties of the uncertainity are made. This Chapter also focuses on the implication of parameter uncertainty, but uses computational techniques. Here the effects of parameter perturbation are used as a metric whereby the controller is evaluated. The metric is not explicitly incorporated into the controller design as in robust control or in the Aoki (1967) method. The focus here is on the condition number of the transfer system matrices in the steady state as the performance metric. For the condition number, the 2-norm is used. So the metric is the ratio of the largest to smallest singular value. Due to the relationship between matrix norms (Burden and Faires, 1985) this choice of the 2-norm is not restrictive.

4.2 Preliminaries

The systems with which this Chapter is concerned are stabilisable, linear, deterministic, time-invariant systems. The aim is to develop a controller for such systems, so that, in the closed loop, they track a desired output trajectory of a set of ramps.

In continuous time, each system is given by

$$\dot{\mathbf{x}}(t) = A\mathbf{x}(t) + B\mathbf{u}(t) + E \tag{4.1}$$

$$\mathbf{y}(t) = C\mathbf{x}(t) + D\mathbf{u}(t) \tag{4.2}$$

where $\mathbf{x}(t) \in \Re^n$ is the state vector, with initial conditions given ($\mathbf{x}(0) = \mathbf{x_0}$); $\mathbf{y}(t) \in \Re^m$ is the output; and $\mathbf{u}(t) \in \Re^m$ is the control vector. The system is square with equal number of inputs and outputs, and thus reflects the Tinbergen (1952) maxim of equal policy instruments and targets. This is not an essential assumption for the analysis. The state space form differs from the usual form used in economic models in that it contains an output equation, and from engineering control models as it contains a vector of intercepts or process levels ($E \in \Re^n$). The advantage of the intercept vector is that it allows for the use of many economic models with little change to the structural form. Both this and the advantage of an output equation were considered in Chapter 3.

The concern of this Chapter is with closed loop systems. Generally, the control rule will be given by

$$\mathbf{u}(t) = \mathbf{r}(t) - K\mathbf{x}(t) \tag{4.3}$$

where $\mathbf{r}(t) \in \Re^m$ is a reference vector. The concern is with $\mathbf{r}(t)$ being a set of ramps. For the Integral control and Linear Quadratic Tracking it is a set of desired outputs; whereas, for the Feedforward control, it is a vector generated by the feedforward controller from a set of desired outputs. K is a feedback gain matrix. For the Integral and Feedforward methods it is, as indicated in Equation 4.3, time invariant; and for the illustrations, is chosen by either pole placement or by a linear quadratic regulator technique. For the Linear Quadratic Tracking method, the feedback gain matrix may be time varying depending on the specification of the objective function. This is considered in Section 4.5.

Multiplier or transfer functions are useful when examining such systems. This is particularly the case when considering the long-run or steady state errors between actual and desired outputs. Consider the situation where the reference vector ($\mathbf{r}$) is a set of desired outputs. For the continuous time system (of Equations 4.1 and 4.2), taking Laplace transforms; assuming zero initial conditions; transforming co-ordinates so intercepts are zero; and closing the loop with Equation 4.3; gives

$$\begin{aligned}\mathbf{Y}(s) &= \{(C - DK)[sI - (A - BK)]^{-1}B + D\}\mathbf{R}(s) \\ &= G(s)\mathbf{R}(s)\end{aligned} \tag{4.4}$$

where s is the Laplace operator; and, $G(s) \in \Re^{m \times m}$. This equation gives the transfer function matrix, $G(s)$, between the actual and desired outputs. From it the multipliers, or input-output relationship, between $Y(s)$ and $R(s)$ can be obtained. Of particular importance are the long-run multipliers given by the Final Value Theorem, for from it, the steady state errors between actual and desired outputs can be determined. The Final Value Theorem gives the long-run outputs by

$$\lim_{t \to \infty} \mathbf{y}(t) = \lim_{s \to 0} s\mathbf{Y}(s) \tag{4.5}$$

The maintained multipliers are given by the unit step response. Here each element of the input vector (in this case $\mathbf{R}(s)$) is s^{-1}, and the Final Value Theorem gives these long-run multipliers as $G(s)$ with $s = 0$. The concern of this Chapter is the response of the model, in the closed loop, to a set of ramp inputs. This can also be given by Equation 4.4 and the Final Value Theorem. In this case the input vector, $\mathbf{R}(s)$, has an ith ($i = 1, 2, \cdots, m$) element of $\alpha_i s^{-2}$ with $\alpha_i \in \Re$. From the Final Value Theorem the steady state response is infinity. This would be expected with the input increasing with time. For this reason particular desired output ramp trajectories are illustrated in this Chapter.

In this Chapter discrete time systems will also be presented; and, again, the same nomenclature will be used for discrete systems, as for continuous systems.The advantage of using the same nomenclature is that it highlights the essentially similar aspects of continuous time and discrete time systems for the analysis here. The nomenclature does not imply that the two systems are the same. With discrete systems, the subscript t will refer to the tth time period, and a discrete system is then written as

$$\mathbf{x}_{t+1} = A\mathbf{x}_t + B\mathbf{u}_t + E \tag{4.6}$$

$$\mathbf{y}_t = C\mathbf{x}_t + D\mathbf{u}_t \tag{4.7}$$

with a control rule of

$$\mathbf{u}_t = \mathbf{r}_t - K\mathbf{x}_t \tag{4.8}$$

The multipliers or transfer function matrix between the actual and desired outputs for the discrete time case is analogous to the continuous time case. With the discrete time system (of Equations 4.6 and 4.7), taking z transforms; assuming zero initial conditions and intercepts; and, closing the loop with Equation 4.8 gives

$$\begin{aligned} \mathbf{Y}(z) &= \{(C - DK)[zI - (A - BK)]^{-1}B + D\}\mathbf{R}(z) \\ &= G(z)\mathbf{R}(z) \end{aligned} \tag{4.9}$$

with $G(z) \in \Re^{m \times m}$. The similar structure of Equations 4.4 and 4.9 show the essentially similar nature for the analysis here of the continuous time and discrete time models.

The Final Value Theorem is given by

$$\lim_{t \to \infty} \mathbf{y}_t = \lim_{z \to 1}[(1 - z^{-1})\mathbf{Y}(z)] \tag{4.10}$$

The unit step and ramp responses are analogous.

Both continuous time and discrete time closed loop systems of this form could have steady state errors to the vector of ramp inputs. In this Chapter methods are considered that overcome these steady state errors.

4.3 Integral Control

4.3.1 Continuous Time

The first method proposed to ensure the system tracks a set of ramps is through the addition of extra integrators to remove the steady state error. The extra integrators are added through the augmentation of the states. Such a method is an extension of that used by Åström and Wittenmark (1990), who considered the single input single output case and aimed to remove the steady state error in tracking a step input. Here it is extended to the multiple input multiple output case, and the aim is to track a ramp for each output.

Suppose the desired output or reference vector ($\mathbf{r}(t)$) contains a set of ramps in the time horizon. Then, after transients, the error between the actual and the desired output will also be a set of ramps while the desired output trajectory is the ramps. Now the second derivative of a ramp trajectory with respect to time is zero. So by double integrating the error, the steady state errors can be removed. The double integrators may be included in the system by augmenting the states.

Consider the errors between the actual output vector and the desired output vector. Then define $\mathbf{e}(t) \in \Re^m$ such that

$$\begin{aligned} \ddot{\mathbf{e}}(t) &= \mathbf{r}(t) - \mathbf{y}(t) \\ &= \mathbf{r}(t) - C\mathbf{x}(t) - D\mathbf{u}(t) \end{aligned} \tag{4.11}$$

Now, as $\ddot{\mathbf{e}} = \mathbf{0}$ with $\mathbf{e}(t)$ being a set of ramps, then $\mathbf{y}(t) = \mathbf{r}(t)$ as desired. If $\mathbf{e}(t)$ is of lower order than a set of ramps (i.e. steps), then its second derivative will also be zero, and so the output vector will equal the desired output vector as required.

Equation 4.11 can be used to augment the system given by Equations 4.1 and 4.2. Firstly, consider each of the individual components of the error vector. The double derivative for each element in $\mathbf{e}(t)$ can be generated as a coupled pair of first order derivatives. This, in turn, can be considered as a small state space system with the error as the first state and its derivative as the second state. The second state is then the output of this system. For the error vector as a whole, the same principle can be used *as long as these states are correctly ordered.*

Using this approach, the system (Equations 4.1 and 4.2) can be augmented by the addition of the error vector and its derivative as the extra states. Define

$$\begin{bmatrix} \mathbf{z}_1(t) \\ \mathbf{z}_2(t) \end{bmatrix} = \begin{bmatrix} \mathbf{e}(t) \\ \dot{\mathbf{e}}(t) \end{bmatrix} \tag{4.12}$$

where $\mathbf{z}_1(t), \mathbf{z}_2(t) \in \Re^m$ are the augmenting states.

The augmenting states can be combined with the original states to produce an augmented state equation of

$$\begin{bmatrix} \dot{\mathbf{x}}(t) \\ \dot{\mathbf{z}}_1(t) \\ \dot{\mathbf{z}}_2(t) \end{bmatrix} = \begin{bmatrix} A_{n\times n} & O_{n\times m} & O_{n\times m} \\ O_{m\times n} & O_{m\times m} & I_{m\times m} \\ -C_{m\times n} & O_{m\times m} & O_{m\times m} \end{bmatrix} \begin{bmatrix} \mathbf{x}(t) \\ \mathbf{z}_1(t) \\ \mathbf{z}_2(t) \end{bmatrix} + \begin{bmatrix} B_{n\times m} \\ O_{m\times m} \\ -D_{m\times m} \end{bmatrix} \mathbf{u}(t) + \begin{bmatrix} E_{m\times 1} \\ O_{m\times 1} \\ O_{m\times 1} \end{bmatrix} + \begin{bmatrix} O_{n\times m} \\ O_{m\times m} \\ I_{m\times m} \end{bmatrix} \mathbf{r}(t) \tag{4.13}$$

The augmented output equation is then given by

$$\mathbf{y}(t) = \begin{bmatrix} C_{m\times n} & O_{m\times m} & O_{m\times m} \end{bmatrix} \begin{bmatrix} \mathbf{x}(t) \\ \mathbf{z}_1(t) \\ \mathbf{z}_2(t) \end{bmatrix} + D_{m\times m}\mathbf{u}(t) \tag{4.14}$$

Where $O_{n\times m}$ is a $n \times m$ matrix of zeros and $I_{n\times n}$ is an $n \times n$ identity matrix. The dimensionality of the parameter matrices is also indicated.

The augmented system can then be written as

$$\dot{\bar{\mathbf{x}}}(t) = \bar{A}\bar{\mathbf{x}}(t) + \bar{B}\mathbf{u}(t) + \bar{E} + \bar{F}\mathbf{r}(t) \tag{4.15}$$

$$\mathbf{y}(t) = \bar{C}\bar{\mathbf{x}}(t) + D\mathbf{u}(t) \tag{4.16}$$

This is the same structure as the original system (Equations 4.1 and 4.2) with the addition of the extra forcing term in the state equation created by the exogenous reference vector. The order of the system has been increased from n to $(n + 2m)$.

The control rule then becomes

$$\mathbf{u}(t) = \mathbf{r}(t) - \bar{K}\bar{\mathbf{x}}(t) \tag{4.17}$$

and the degree of emphasis on the removal of the steady state error can be determined by the choice of the gains in $\bar{K}$. This could be done, for example, by using a linear quadratic regulator.

The strength of this method lies in the fact that it simply augments the original system with matrix partitions generated by the addition of the extra states. This occurs because of the ordering of the additional states. The augmenting states are considered as a vector of errors followed by a vector of error derivatives, rather than as, for each ramp input, an error followed by its derivative. Ordering the augmenting states in this manner means that adding additional ramp inputs simply alters the dimensionality of the augmenting states but not their structure.

Another advantage of this method is that it extends up to higher order polynomial inputs. The above augmented system will track inputs of polynomials in time of degree less than two (i.e. steps and ramps) with no steady state errors. It will not track higher degree polynomials as their second derivative is not zero. Yet the method can be scaled up so that, for example, triple integrators can be used for inputs that are quadratics in time, for here the third derivative is zero. Triple integrators can be included using the same approach but require another block of m states to be included in the augmented system. The same argument applies to higher degree polynomial inputs. Polynomial inputs of degree p can be tracked with no steady state errors by using this method and augmenting the system with pm additional states, as long as the additional states are ordered in the correct manner.

A disadvantage of the technique is that integrators can lead to stability problems. The higher the degree of the polynomial input, the more integrators and the more likely stability problems will be introduced.

The question arises as to whether the augmentation alters other significant properties held by the original system. Consider the augmentation for a set of ramp inputs as given in Equations 4.13 and 4.14. Notice that the augmentation has added a block of m columns of zeros to the state equation driving matrix, and thus it will not be of full rank even if the original state equation driving matrix was. Of more importance is the rank of the controllability matrix of the augmented system. Specifically, an important question is: if the original system is controllable, then is the augmented system also controllable? The augmented system is completely state controllable if the rank of its controllability matrix

$(\mathcal{C}_{\bar{A}\bar{B}})$ is equal to the system order. That is, the rank of $\mathcal{C}_{\bar{A}\bar{B}}$ is $\bar{n} = n + 2m$, where $\mathcal{C}_{\bar{A}\bar{B}}$ has dimensionality of $\bar{n} \times \bar{n}m$. Now

$$\mathcal{C}_{\bar{A}\bar{B}} = [\,\bar{B} \quad \bar{A}\bar{B} \quad \bar{A}^2\bar{B} \quad \bar{A}^3\bar{B} \quad \cdots \quad \bar{A}^{\bar{n}-1}\bar{B}\,]$$
$$= \begin{bmatrix} B & AB & A^2B & A^2B & \cdots & A^{\bar{n}-1}B \\ O & -D & -CB & -CAB & \cdots & -CA^{\bar{n}-3}B \\ -D & -CB & -CAB & -CA^2B & \cdots & -CA^{\bar{n}-2}B \end{bmatrix} \tag{4.18}$$

where O is an $m \times m$ zero matrix.

Equation 4.18 is the controllability matrix between $\bar{A}$ and $\bar{B}$ in Equation 4.15. Equation 4.15 also has another forcing function of the reference, $\mathbf{r}$. Now this reference vector can be considered as a control, and could be generated, like in the Feedforward approach below, to give the closed loop system desirable characteristics. Specifically $\mathbf{r}$ could be considered as the 'overall' control for the system, and, then, could be used to drive the states to a given point in state space. That is, it is actually $\mathbf{r}$ that is the important forcing function, and the important controllability implications are from $\mathbf{r}$ to the augmented states $\bar{\mathbf{x}}$.

Closing the loop by substituting Equation 4.17 into Equation 4.15, gives

$$\dot{\bar{\mathbf{x}}}(t) = (\bar{A} - \bar{B}\bar{K})\bar{\mathbf{x}}(t) + (\bar{B}\bar{F})\mathbf{r}(t) + \bar{E} \tag{4.19}$$
$$= \bar{\bar{A}}\bar{\mathbf{x}}(t) + \bar{\bar{B}}\mathbf{r}(t) + \bar{E} \tag{4.20}$$

It is the controllability of $(\bar{\bar{A}}, \bar{\bar{B}})$ that is important, that is the rank of $\mathcal{C}_{\bar{\bar{A}}\bar{\bar{B}}}$.

The problem with this argument is that it assumes the state feedback gain matrix, $\bar{K}$, is given. If $(\bar{A}, \bar{B})$ is not controllable, then $\bar{K}$ cannot be obtained. For the required relationship, in the closed loop, between the reference and the states, requires that the controls can be such that they can drive the states to any position in the state space. Thus for $(\bar{\bar{A}}, \bar{\bar{B}})$ to be controllable, $(\bar{A}, \bar{B})$ must be controllable. Thus Equation 4.18 gives the appropriate controllability matrix.

Now consider the effects of augmenting the states on the controllability of the augmented system. In a comparison between the original system's controllability matrix, $\mathcal{C}_{AB}$, and that of the augmented system, $\mathcal{C}_{\bar{A}\bar{B}}$, it can be seen that: firstly, the augmentation has significantly added to the dimensionality; and, secondly, the original system's output matrices are introduced into the calculation. The higher dimensionality increases the risk of numeric problems with evaluating Equation 4.18; and, it is frequently the case that the output matrices are sparse; but it is not immediately obvious from Equation 4.18 whether the augmentation process will not fully add to the rank of the controllability matrix. A simple example shows that it does not.

As an example consider the Currie and Levine (1993*a*) model given in Chapter 2 (Section 2.5 on page 36). In their state space form this is a fourth order model but the state equation driving matrix contains a row of zeros, and will not be of full rank, (as can be seen from Equation 2.53 on page 38). Using the same parameter values as Currie and Levine (1993*a*), the original system is controllable and the augmented system is controllable. In this form, the model is used to illustrate the Integral method in Section 4.3.2 (on page 122) below. Modifying the parameter ϕ_1 from one to zero modifies the state equation control matrix, B. This modification leaves the system still being controllable with the rank of $\mathcal{C}_{AB}$ the same as the system size (4). The actual values of the state equation driving matrix, A; the original state equation control matrix, B_o; and the modified matrix, B_m, are

$$A = \begin{bmatrix} -0.4925 & 0.5 & -0.1625 & 0.1625 \\ -1.45 & 0 & 0.15 & -0.15 \\ 0.15 & 0 & -0.25 & 0.25 \\ 0 & 0 & 0 & 0 \end{bmatrix} \qquad B_o = \begin{bmatrix} -0.05 & -0.2 \\ 0 & -1 \\ 0 & 0 \\ 1 & 0 \end{bmatrix} \qquad B_m = \begin{bmatrix} -0.05 & -0.2 \\ 0 & 0 \\ 0 & 0 \\ 1 & 0 \end{bmatrix}$$

Now consider augmenting the modified system. Augmentation increases the system size to 8, but the rank of the augmented controllability matrix, $\mathcal{C}_{\bar{A}\bar{B}}$, is only increased to 7. That is, the original system is controllable, but the augmented system is not. Thus augmenting the system lost the important property of controllability for this modified system. This example shows that the important property of controllability can be lost through augmenting the system with additional integrators to track a ramp.

It is interesting to note that augmenting this modified Currie and Levine (1993*a*) system to track a step in its inputs also looses the property of controllability. To track a ramp the system had to be augmented with two double integrators, as a double integrator is necessary for each individual output. In the step case a pair of single integrators (one for each of the two outputs) is all that is needed. This is because the first derivative of a step is zero, so that the augmenting states need only include the first derivative. Then, in Equation 4.11, $\ddot{\mathbf{e}}$ can be replaced with $\dot{\mathbf{e}}$ resulting in the system being augmented by only m additional states. For this modified Currie and Levine (1993*a*) system, the augmented system size will now be 6, but the rank of this augmented system's controllability matrix is only 5. Thus augmenting the modified system to only track a step, also looses the property of controllability (c.f. Ogata, 1997, p. 887).

Controllability is an important property of the state space systems used here. Another important property is that of observability. This book makes much use of observers, so that it is important that the states can be re-created from the outputs. Adding integrators through the augmentation process will not add to the rank of the observability matrix for the augmented system, so the augmented system will be unobservable. This is expected as it would seem unreasonable to generate the errors from the outputs. Now, the observability matrix for the augmented system is

$$\mathcal{O}_{\bar{A}\bar{B}} = \begin{bmatrix} \bar{C} \\ \bar{C}\bar{A} \\ \bar{C}\bar{A}^2 \\ \bar{C}\bar{A}^3 \\ \vdots \\ \bar{C}\bar{A}^{\bar{n}-1} \end{bmatrix} = \begin{bmatrix} C & O \\ CA & O \\ CA^2 & O \\ CA^3 & O \\ \vdots & \\ CA^{\bar{n}-1} & O \end{bmatrix} \tag{4.21}$$

where O is, for the augmented system to track a set of ramp inputs, an $n \times 2m$ zero matrix. It can be seen from Equation 4.21 that the augmentation will not add to the rank of the observability matrix. Transforming the system into observable canonical form (as considered on page 91) may allow the system to be observable, but a tradeoff may be (the unexceptable) lack of controllability.

The lack of observability of the augmented system is not such a problem here where state feedback is used. But the concept of observability is an important one for this book, for observers are used extensively. In Chapter 5 (Section 5.3 on page 161), the augmented system is used with an observer despite the lack of observability. This is not a problem as the states associated with the errors go to zero, as is the aim of Equation 4.11.

Thus one problem with the addition of integrators through augmentation is that it could loose the property of controllability. Adding integrators through other means than augmenting the states may be used to overcome this. This is considered in Chapter 5 (Section 5.3.2 on page 166).

Another disadvantage of the augmentation method is that it adds integrators into the system. The addition of integrators can create instability in the system. Though, in the closed loop, this should not be a problem.

The augmentation process also adds m columns of zeros into the state equation driving matrix ($\bar{A}$). It will thus have a smallest singular value using the 2-norm of zero, so that the matrix will be ill-conditioned with its condition number (using the 2-norm) being infinite ($\kappa_2(\bar{A}) = \infty$). The matrix will be significantly effected by parameter perturbations. Here the focus is more on the effects of perturbations of the state equation driving matrix on its eigenvalues. In this case the condition number of the matrix is not the important metric, but the condition number of its eigenvector matrix. The eigenvector matrix need not be ill-condition just because the original matrix is ill-conditioned (Mathworks, 1992, p.64).

Of more interest, here, is the closed loop transfer function matrix which gives the relationship between the reference vector, $\mathbf{r}(t)$, and the output vector, $\mathbf{y}(t)$. That is, the augmented system's multipliers. It is the robustness of this relationship to parameter perturbations that is of greatest interest. Again, taking Laplace transforms; assuming zero initial condi-

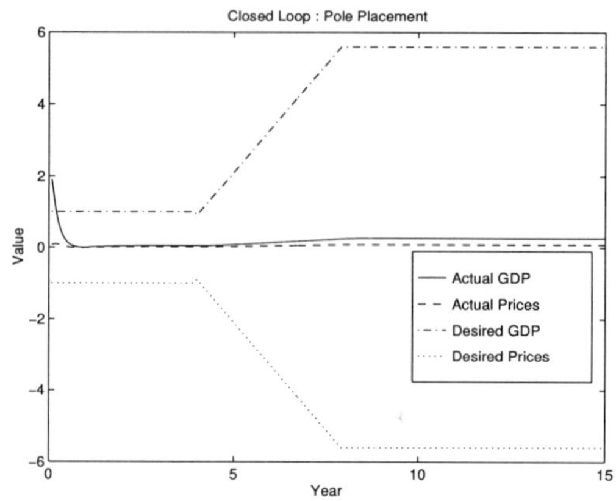

Figure 4.1: Continuous Model: Closed Loop

tions for Equations 4.15 and 4.16; and, closing the loop with Equation 4.17, gives

$$\begin{aligned}\mathbf{Y}(s) &= \{(\bar{C} - D\bar{K})[sI - (\bar{A} - \bar{B}\bar{K})]^{-1}(\bar{B} + \bar{F}) + D\}\mathbf{R}(s) \\ &\quad + (\bar{C} - D\bar{K})[sI - (\bar{A} - \bar{B}\bar{K})]^{-1}E \\ &= \bar{G}(s)\mathbf{R}(s) + \bar{N}\end{aligned} \tag{4.22}$$

where s is the Laplace operator; $\bar{G}(s) \in \Re^{m \times m}$; and, $\bar{N} \in \Re^{m}$.

The condition number of the transfer function matrix ($\bar{G}(s)$), and the condition number of its eigenvector matrix are metrics that can be used to consider the robustness of the system to parameter perturbations.

4.3.2 Continuous Time Example

For an example of this method, the deterministic Currie and Levine (1993*a*) unmodified model referred to above in Section 4.3.1 (on page 119; and in greater detail in Section 2.5 on page 36) will be used. This model has the two inputs of interest rates and autonomous taxes; and the two outputs of GDP and the price level. The state form of the model is conditionally stable with the state equation driving matrix (A) having a row of zeros. With the parameter values used by Currie and Levine (1993*a*), the system is both controllable and observable. The system size is 4 and rank of $\mathcal{C}_{AB}$ and the rank of $\mathcal{O}_{AC}$ are both 4.

The open loop[1] model is stable to a unit step in autonomous taxes, but the price level is unstable to a unit step in interest rates. In the open loop, then, the system will be unable to track a set of ramp trajectories in desired output. The closed loop response to tracking desired output trajectories[2] is given in Figure 4.1. The desired output contains an initial step followed by a ramp to a plateau. The desired response for each of the individual outputs is a mirror image of the other. The state feedback gain matrix, K, was chosen using the pole placement technique with the closed loop poles equally spaced between -2.5 and -5. The Figure plots the desired trajectories and the actual trajectories, and it shows that, in this closed loop system, the desired outputs are not attained. There is a large steady state error. This illustrates how pole placement is concerned with the placement of eigenvalues and thus the stability of the states, rather than upon desired outputs.

Augmenting the system through the addition of double integrators for each of the two outputs doubles the system size to 8. The system is still state controllable as the rank of the augmented controllability matrix ($\mathcal{C}_{\bar{A}\bar{B}}$) is 8, but not observable as the rank of the observability matrix is not increased.

The tracking of a set of ramps with this augmented system is presented in Figures 4.2 and 4.3. Two control techniques have been used to determine the feedback gain matrix ($\bar{K}$), with the first being pole placement and the second being the linear quadratic regulator. In the latter case the weights chosen are: unity for controls; 100 for the original states; and, 10000 for the augmenting states. The weights have been chosen so that the linear quadratic regulator has closed loop poles of the same order of magnitude as that generated by the pole placement technique. This can be seen from Table 4.1 where the eigenvalues of the closed loop state equation driving matrix are presented. One of the features of Linear Quadratic

[1] All simulations are undertaken using Euler numerical technique with a step size of one month.

[2] The ramps are for illustrative purposes and are not intended to make economic sense with the model.

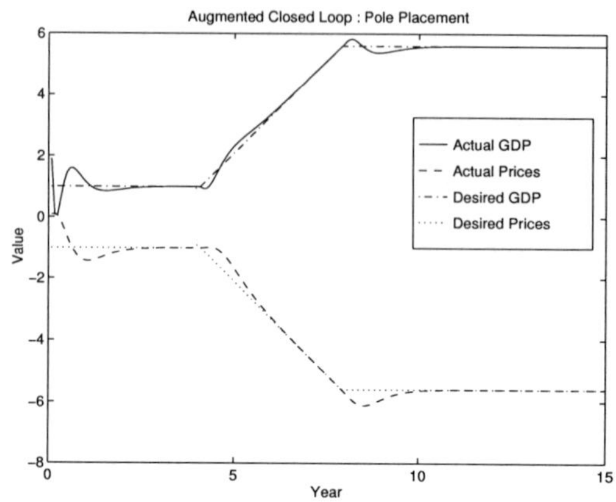

Figure 4.2: Continuous Model: Integral Method Using Pole Placement

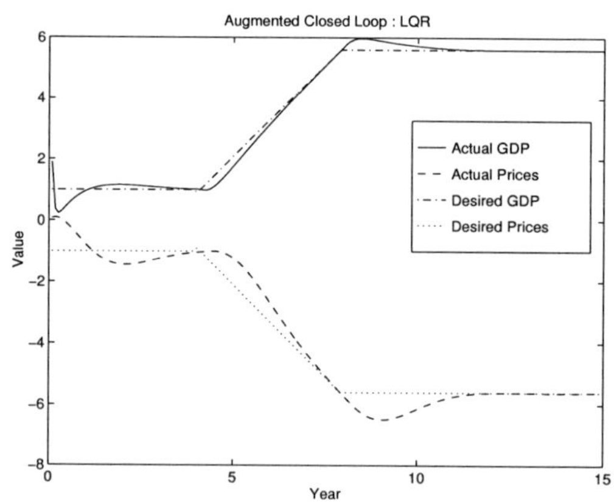

Figure 4.3: Continuous Model: Integral Method Using LQR

Eigenvalues of $(\bar{A} - \bar{B}\bar{K})$	
Pole Placement	LQR
−5.0000	$-9.9690 + 0.0185\imath$
−4.6429	$-9.9690 - 0.0185\imath$
−2.5000	$-1.8719 + 1.3875\imath$
−4.2857	$-1.8719 - 1.3875\imath$
−2.8571	$-1.0308 + 1.1941\imath$
−3.9286	$-1.0308 - 1.1941\imath$
−3.2143	−0.9850
−3.5714	−0.9452

Table 4.1: Continuous System: Eigenvalues of Closed Loop State Equation Driving Matrix

Regulator design is that the choice of actual pole placement is removed from the design. It is thus difficult to obtain given closed loop poles using Linear Quadratic Regulator design.

Figures 4.2 and 4.3 show that the Integral control method introduced through the augmentation of the states forces the system to track the desired outputs of a ramp and a step with no steady state error. Notice that this control strategy acts *ex post* to the changing desired output. It is a reactive strategy.

It is obvious from a comparison of the results of the closed loop augmented system (Figures 4.2 and 4.3) with the original closed loop system (Figure 4.1), that this control strategy significantly improves the ability to track desired outputs containing sets of ramps.

For a comparison between the control techniques, and for a comparison with the other methods considered below, Table 4.1 contains the eigenvalues of the closed loop state equation driving matrix. While the eigenvalues of the closed loop state equation driving matrix from the different techniques of choosing the feedback gain are not the same, they give an indication of the speed of the response of the augmented closed loop system. They also give an indication of the speed of the various methods of tracking a set of ramps.

Also, for a comparison with other techniques, the tracking error is measured as the relative error between the desired output and the actual output (using the 2-norm). Again, this gives an indication of the tracking ability of the various methods. In this illustration, the tracking error using the pole placement technique is 0.04, and for the linear quadratic regulator is 0.08.

For the implications of parameter perturbations consider the transfer system matrix in the long run. This is given by $\lim_{s\to 0}\bar{G}(s)$ (from Equation 4.22). Call this matrix $\bar{G}$. This matrix is well-conditioned both when the pole placement technique is used ($\kappa_2(\bar{G}_{pp}) = 1.2$), and with the linear quadratic regulator ($\kappa_2(\bar{G}_{lq}) = 1.3$). The system is then robust to small parameter perturbations. For the effects on the eigenvalues of the transfer system matrix, then the relevant matrix is that whose columns are the eigenvectors of $\bar{G}$. For the pole placement technique this matrix has a condition number of 2.6; and, for the linear quadratic regulator, its condition number is 1.7. Again this indicates that it is well-conditioned to small parameter perturbation. It compares to the infinitely ill-conditioned open loop situation ($\kappa_2(A) = \infty$).

4.3.3 Discrete Time

The same approach can be used for discrete systems, and the derivation is analogous. Again the system is augmented by including extra states based on the appropriately degreed derivative of the error vector, but a discrete approximation is used for the derivative. As long as the augmenting states are ordered in the correct way, then, for the augmentation, it does not matter how many outputs there are, for the system matrices are simply block augmented. Similarly, the number of sets of augmenting states, and thus augmenting system matrices blocks, depends upon the degree of the polynomial (in time) of the desired output vector trajectory.

Here the emphasis is on tracking sets of ramps for the desired output trajectories; and, so, double integrators are appropriate, for the second derivative of the error vector will be zero. Thus the system needs to be augmented with twice as many states as there are outputs to track desired output trajectories of ramps. Again, ordering these augmenting states in the appropriate way allows for the expanding of the system matrices with blocks of identity and zero matrices.

The basic approach proposed here, for discrete systems, is to approximate the double derivative using a central difference method, and to combine this approximation into the sumers for the error between the desired and actual outputs. In this way the truncation error caused by the approximation is amalgamated with the tracking error.

In the scalar case, a central difference approximation (Burden and Faires, 1985) for the double derivative of an error (e) with respect to time (t) is given by

$$\ddot{e}(t) = \frac{e(t_{i+\Delta t}) - 2e(t_i) + e(t_{i-\Delta t})}{(\Delta t)^2} + o(\Delta t)^2 \tag{4.23}$$

where $o(n)$ refers to order n, and Δt is the step size. Using the vector analogue of Equation 4.23, then the tracking error equation, Equation 4.11 (on page 115), can be approximated *at the discrete time steps* by

$$\mathbf{e}_{t+1} = \mathbf{r}_t - C\mathbf{x}_t - D\mathbf{u}_t + 2\mathbf{e}_t - \mathbf{e}_{t-1} \tag{4.24}$$

Equation 4.24 not only includes the tracking error but also the truncation error associated with approximating the double derivative at the grid points. This error will also be 'integrated out' in the steady state.

Define the augmenting states as

$$\begin{bmatrix} \mathbf{z}_{1,t} \\ \mathbf{z}_{2,t} \end{bmatrix} = \begin{bmatrix} \mathbf{e}_{t-1} \\ \mathbf{e}_t \end{bmatrix} \tag{4.25}$$

Then augmented system state equation becomes

$$\begin{aligned} \begin{bmatrix} \mathbf{x}_{t+1} \\ \mathbf{z}_{1,t+1} \\ \mathbf{z}_{2,t+1} \end{bmatrix} &= \begin{bmatrix} A_{n\times n} & O_{n\times m} & O_{n\times m} \\ O_{m\times n} & O_{m\times m} & I_{m\times m} \\ -C_{m\times n} & -I_{m\times m} & 2I_{m\times m} \end{bmatrix} \begin{bmatrix} \mathbf{x}_t \\ \mathbf{z}_{1,t} \\ \mathbf{z}_{2,t} \end{bmatrix} + \begin{bmatrix} B_{n\times m} \\ O_{m\times m} \\ -D_{m\times m} \end{bmatrix} \mathbf{u}_t \\ &+ \begin{bmatrix} E_{m\times 1} \\ O_{m\times 1} \\ O_{m\times 1} \end{bmatrix} + \begin{bmatrix} O_{n\times m} \\ O_{m\times m} \\ I_{m\times m} \end{bmatrix} \mathbf{r}_t \end{aligned} \tag{4.26}$$

The augmented system output equation is

$$\mathbf{y}_t = \begin{bmatrix} C_{m\times n} & O_{m\times m} & O_{m\times m} \end{bmatrix} \begin{bmatrix} \mathbf{x}_t \\ \mathbf{z}_{1,t} \\ \mathbf{z}_{2,t} \end{bmatrix} + D_{m\times p}\mathbf{u}_t \tag{4.27}$$

The augmented system can, again, be written as

$$\bar{\mathbf{x}}_{t+1} = \bar{A}\bar{\mathbf{x}}_t + \bar{B}\mathbf{u}_t + \bar{E} + \bar{F}\mathbf{r}_t \tag{4.28}$$

$$\mathbf{y}_t = \bar{C}\bar{\mathbf{x}}_t + D\mathbf{u}_t \tag{4.29}$$

The augmentation of a discrete system to track a set of ramps for the desired output is thus analogous to the augmentation of a continuous time system. The augmented system matrices are similar. The main difference is in the state equation driving matrix, $\bar{A}$, which has a different block pattern in its lower right hand corner. The matrix for the discrete time case differs from its continuous analogue in that it does not have a column block (of width m) of zeros. It can thus be of full rank and need not be infinitely ill-conditioned.

As with the continuous time analogue, augmenting the discrete time system may lose desirable properties. If the original discrete time system is controllable then it does not follow that the augmented system will also be controllable. This is shown by an example in the illustration of the discrete time augmentation to track a set of ramps in Section 4.3.4 (on page 131) below.

The observability matrix for the augmented discrete time case will have the same format as its continuous analogue (Equation 4.21). The augmented discrete system will, too, be unobservable.

The transfer system matrix for the discrete system will be the same format as in the continuous case (Equation 4.22). Though z transforms will be used in the discrete time case rather than the Laplace transforms used in the continuous time case. The metrics are analogous.

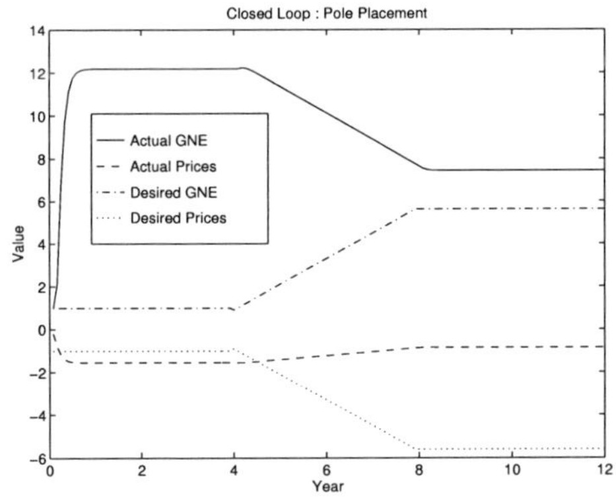

Figure 4.4: Discrete Model: Closed Loop

4.3.4 Discrete Time Example

For an example of a discrete application of the Integral control method, the three state discrete time model developed in Chapter 3 (in Section 3.3.5 on page 75) is used. This model has the two inputs of a fiscal policy and a monetary policy instrument; and, two outputs of GNE[3] and the price level. The model, in the open loop, is slow and stable. It is also both controllable and observable with the rank of both $\mathcal{C}_{AB}$ and $\mathcal{O}_{AC}$ being the same as the system size of 3.

The closed loop response of the system to a desired output trajectory is given visually in Figure 4.4. Here the feedback gain matrix (K) is chosen by the pole placement technique and poles are equally spaced between 0.25 and 0.4. The Figure plots the desired output trajectory and the actual output trajectory. It clearly shows that, in this system, the desired output trajectory is far from being attained; and illustrates, again, how pole placement, in general, does not result in attaining desired outputs.

Augmenting the system through the addition of sumers for each of the two outputs increases

[3] For scaling purposes GNE is measured in tens of millions of dollars. Specifically, the initial value of GNE is divided by 10 in this illustrative example.

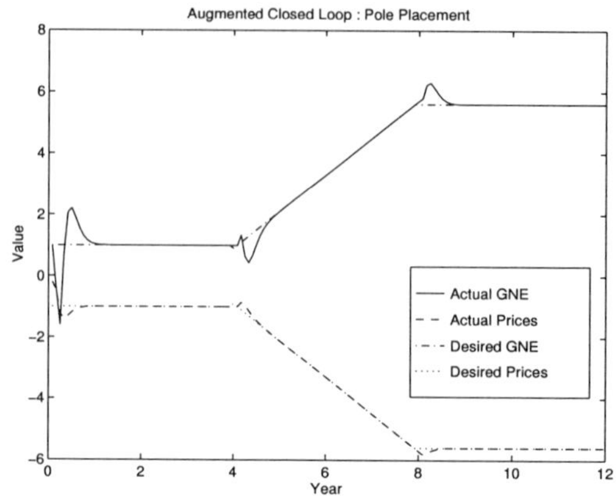

Figure 4.5: Discrete Model: Integral Method Using Pole Placement

the system size to 7. The system is still state controllable as the rank of the augmented controllability matrix ($\mathcal{C}_{\bar{A}\bar{B}}$) is 7, but not observable as the rank of the observability matrix is not increased.

The tracking of a set of ramps with this augmented system is presented in Figures 4.5 and 4.6. Again, pole placement and linear quadratic regulator control techniques are used. The linear quadratic regulator weights are: unity for controls and for the augmenting states; and, 100 for the original states. The pole placement technique places the poles equally spaced between 0.25 and 0.4. Again, these values were chosen to be illustrative and to produce closed loop systems with approximately similar poles. Table 4.2 contains the eigenvalues of the closed loop state equation driving matrix.

The Figures show how the augmented system tracks the desired output trajectories. This is the situation for the initial steps, the ramps and the constants (flats) in desired output. A comparison between the augmented system (Figures 4.5 and 4.6) and the original system (Figure 4.4) shows the enhanced performance of this control strategy in tracking desired output trajectories that contain a set of steps and ramps. For the augmented system, the tracking error using the pole placement technique is 0.06, and for the linear quadratic

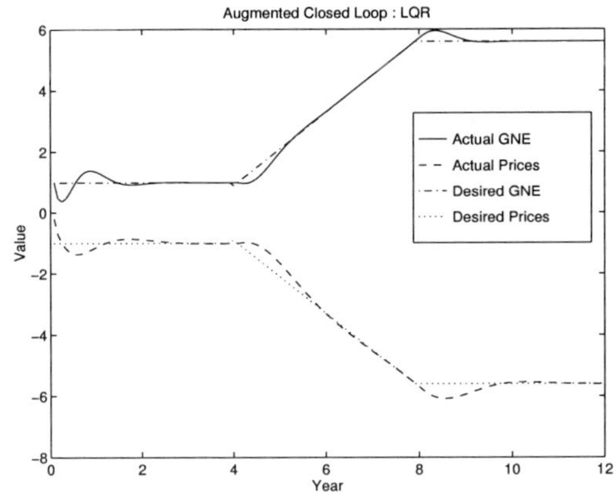

Figure 4.6: Discrete Model: Integral Method Using LQR

regulator is 0.04. This compares to a tracking error of 1.38 for the original closed loop system.

For this discrete case, the transfer system matrix in the long run, $\bar{G}$, is approximately the identity matrix for both control techniques. Thus the matrices are very well-conditioned. For the pole placement technique the matrix made of columns of the eigenvectors of $\bar{G}$ has a condition number of 11.1 and is unity for the linear quadratic regulator. Again the eigenvalues are well-conditioned to small parameter perturbation.

Eigenvalues of $(\bar{A} - \bar{B}\bar{K})$	
Pole Placement	LQR
0.2500	0.0120
0.2750	0.7024
0.3000	0.8162
0.3250	$0.7988 + 0.2464\imath$
0.3500	$0.7988 - 0.2464\imath$
0.3750	$0.8869 + 0.1520\imath$
0.4000	$0.8869 - 0.1520\imath$

Table 4.2: Discrete System: Eigenvalues of Closed Loop State Equation Driving Matrix

The model may be modified to illustrate how the property of controllability can be lost through the augmentation process. Modifying the state equation control forcing matrix, B, so that the final element in the bottom row is zero ($B_{3,2} = 0$), leaves the system still controllable and observable. When the system is augmented to track the set of ramps, the system size is, again, increased to 7, but the rank of the augmented system's controllability matrix is only increased to 5. Thus with this modified system, the original system is controllable, but the augmented system is not.

4.4 Feedforward Control

The Feedforward method uses a different approach to the Integral method. Rather than focusing on the model, the Feedforward method focuses on the desired output, and attempts to generate a reference for use in the control rule so that, in the closed loop steady state, the model generates an output that is the desired output. The basic strategy of generating this reference is through model inversion. The method uses the fact that the closed loop steady state dynamic properties of the model are known for the class of inputs of concern (i.e. sets of ramps and steps).

As with the Integral method, the continuous time and discrete time cases are analogous; and the method will be developed and illustrated for both these cases.

4.4.1 Continuous Time

Consider the closed loop continuous time system presented in Section 4.2 above (on page 111). That is the system given by the state equation (Equation 4.1) and output equation (Equation 4.2), together with the control rule (Equation 4.3). Now in the closed loop and after the transients, the states will grow at a constant rate to a set of ramp inputs. So, in the steady state to the reference ($\mathbf{r}$) of a set of ramps,

$$\dot{\mathbf{x}}(t) = (A - BK)\mathbf{x}(t) + E + B\mathbf{r}(t) = \gamma \tag{4.30}$$

where $\gamma \in \Re^n$ (with n being the system size). If $\mathbf{r}(t)$ is constant, that is a set of steps, then γ will be the zero vector, for then the states will be stationary in the steady state.

The closed loop steady state dynamic properties of the system are given by Equation 4.30, and it can be inverted to generate the reference necessary for the system to produce the desired output.

From Equation 4.30 and Equation 4.2 the states and hence the output can be obtained. That is,

$$\mathbf{y}(t) = (C - DK)(A - BK)^{-1}[\gamma - B\mathbf{r}(t) - E] + D\mathbf{r}(t) \tag{4.31}$$

Then, given a desired output ($\mathbf{y_d}$) of a set of ramps and steps, Equation 4.31 can be inverted to generate the reference necessary to produce this output. That is,

$$\mathbf{r}(t) = (D - MB)^{-1}[\mathbf{y_d}(t) - M(\gamma - E)] \tag{4.32}$$

where

$$M = (C - DK)(A - BK)^{-1} \tag{4.33}$$

Thus, after the transients, a reference given by Equation 4.32 when used in the control rule (Equation 4.3) will force the output (Equation 4.2) to track the desired trajectory.

Firstly, note that the matrix inverses, necessary to generate this reference, will exist since the feedback gain (K) has been chosen to stabilise the system. That is, the model can be inverted.

There is a problem, though, in that the state growth rates (γ) are not known in the case of ramp inputs. In the case of $\mathbf{y_d}$ being a set of steps, than $\mathbf{r}$ can be found analytically as $\gamma = \mathbf{0}$. For the ramp case, the states growth rates may be found by the numerical technique of fixed point iteration (Burden and Faires, 1985). Iterating Equation 4.30 from initial conditions with $\mathbf{r}(t) = \mathbf{y_d}(t)$ for $\mathbf{y_d}(t)$ being a set of ramps, will converge to a fixed point. Convergence is guaranteed as, again, the feedback gain (K) is chosen to stabilise the system.

Thus a numerical procedure can be used to find the states growth rate. This procedure requires *ex ante* knowledge of the desired ramp output. It can, though, be undertaken independently of the actual control implementation (off-line).

The generation of the reference is outside the loop. The control rule can be given by substituting Equation 4.32 into Equation 4.3, giving

$$\mathbf{u}(t) = (D - MB)^{-1}[\mathbf{y_d}(t) - M(\gamma - E)] - K\mathbf{x}(t) \tag{4.34}$$

It can be bought inside the loop, giving the control rule of

$$\mathbf{u}(t) = (D - MB)^{-1}[\mathbf{y_d}(t) - M(\gamma - E) - (D - MB)K\mathbf{x}(t)] \tag{4.35}$$

This latter rule may give other desirable closed loop properties including reduced sensitivity to parameter perturbation.

The Integral method has the advantage over the Feedforward method in that it does not require advanced knowledge of the reference. The Integral method also has the advantage that it can be extended to allow for higher order polynomial in time references rather than steps and ramps. On the other hand, integrators can lead to instability. The Feedforward method does not add the integrators. Another advantage of the Feedforward method is that it does not effect the controllability or observability of the system. It does, though, involve a numerical technique, and a number of matrix inversions, and these could be sensitive to numerical problems. The Integral is also more robust to numerical problems as it is designed to 'integrate out' errors.

4.4.2 Discrete Time

The discrete time Feedforward method is analogous to the continuous time Feedforward method; with, again, the approach being to find the closed loop steady state values for the states, and, to invert the system to generate a reference that produces the desired output.

For the discrete time system given by the state equation of Equation 4.6, the output equation of Equation 4.7 and the loop closed with Equation 4.8, then, after the transients, the states will evolve to a set of ramp inputs according to

$$\mathbf{x}_{t+1} = W\mathbf{x}_t \tag{4.36}$$

where $W \in \Re^{n \times n}$ is a diagonal matrix. For a set of step inputs, W will be an identity matrix, for the states will be constant in the steady state.

Preceding as in the continuous case above, then from Equation 4.36 the states and hence the output can be obtained. That is,

$$\mathbf{y}_t = (C - DK)(W - (A - BK))^{-1}[B\mathbf{r}_t + E] + D\mathbf{r}_t \tag{4.37}$$

So, model inversion leads to

$$\mathbf{r}_t = (D + \mathcal{M}B)^{-1}(\mathbf{y}_{\mathbf{d},t} - \mathcal{M}E) \tag{4.38}$$

where $\mathbf{y_d}$ is the desired output, and

$$\mathcal{M} = (C - DK)[W - (A - BK)]^{-1} \tag{4.39}$$

This is the discrete analogue of Equation 4.32, and can be used to determine the reference that generates the desired output that are sets of ramps. Steps and constants can also be included in the desired output trajectory, but not higher order polynomials in time.

There is also the problem of finding the diagonal matrix W for the set of ramp inputs. The matrix contains the closed loop steady state growth rate for each state as the corresponding diagonal entry. Again, it can be found numerically by fixed point iteration of the closed loop state equation using $\mathbf{r}_t$ of the desired output values. Convergence is guaranteed as the state feedback gain matrix, K, is chosen to stabilise the system. Specifically, let ω_{t+1} be a vector containing the growth rate for each state between successive time periods, that is from t to $t+1$. Then the closed loop state equation and an equation for ω_{t+1} can be iterated until the relative error between successive iterates of ω is less than some tolerance. Then W is found from the fixed point of ω plus the identity matrix.

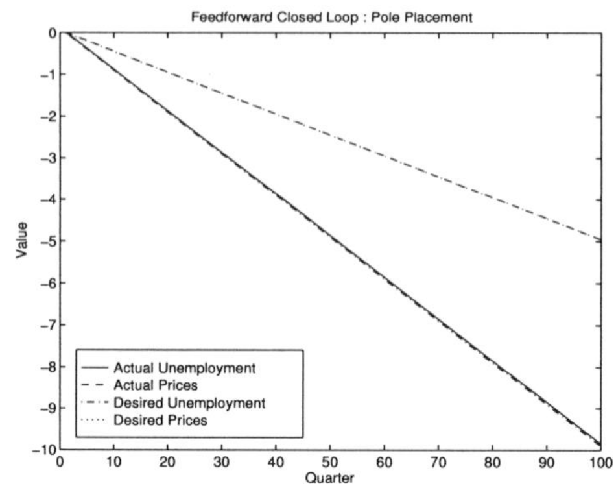

Figure 4.7: Feedforward Method

The discrete time Feedforward method is subject to the same advantages and disadvantages as its continuous time analogue. These include that it requires *ex ante* knowledge of the desired outputs, and it could be susceptible to numeric problems.

4.4.3 Feedforward Examples

Three illustrations of the Feedforward method are given; with the first illustrating the method being used to track a set of ramps, and the other two are comparable with the illustrations used for the Integral method. The latter illustrations allow some comparison of the different methods.

The first illustration uses the stationary contemporaneous control model in Chapter 3 (in Section 3.4.4 on page 98). This is a discrete time, eight state model with the two outputs of the inflation rate and the unemployment rate and two inputs of a fiscal policy instrument and a monetary policy instrument. The model is estimated using Quarterly Australian data. It is unstable in the open loop.

The simulation scenario is as follows: initial state values are zero (for, recall, that this model has been normalised); and the aim is to track a set of ramps, with that for both the inflation rate and the unemployment rate decreasing. The ramps have different rates, with desired inflation decreasing at twice the rate of unemployment. The characteristics of the simulation are: firstly, pole placement is used to determine the feedback gain matrix, and the poles have been chosen to be equally spaced between 0.05 and 0.08, making a fast response; secondly, the fixed point iterative scheme is run out over the same time horizon as the model, and is separate to the actual control; and, thirdly, the reference calculation is outside the loop with Equation 4.34 being used as the control rule. For the simulations here, the calculation of the reference outside or inside the loop makes no difference to the results.

The results of the simulation are shown visually in Figure 4.7. The Figure shows that the method ensures that the ramp trajectories are tracked well. The tracking error is 0.007.

The second illustration uses the same continuous time model as was used to illustrate the continuous time Integral method. It uses the same scenario, including the same desired output trajectories. Again, both pole placement and linear quadratic regulator techniques are used to determine the state feedback gain matrix. The illustration is comparable to that used above in Section 4.3.2 (on page 122) for the Integral method.

For the fixed point iterative procedure, the closed loop state equation, Equation 4.30, is run out using an Euler approximation until the relative error (using the 2-norm) between successive state iterates is small ($< 1 \times 10^{-6}$). This equation is driven by a reference of the desired output.

The results are presented visually in Figures 4.8 and 4.9. Figure 4.8 uses pole placement technique with eigenvalues of the closed loop state equation driving matrix placed as given in Table 4.3. The linear quadratic regulator technique is used for the results presented in Figure 4.9. Here weights are unity for controls, 1000 for the output states and 100 for the other states. The resulting eigenvalues for the state equation driving matrix are also given in Table 4.3. Again, the values are chosen to facilitate comparison amongst the methods.

The Figures show a poorer performance than with the Integral method. This can be seen from comparing Figure 4.8 with Figure 4.2 (on page 124); and comparing Figure 4.9 with

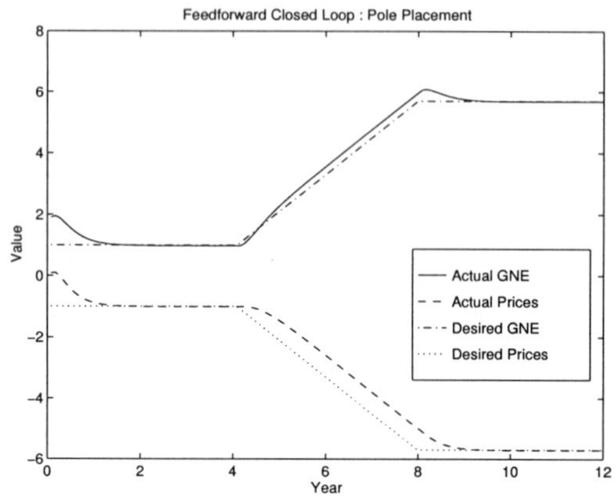

Figure 4.8: Continuous Model: Feedforward Method Using Pole Placement

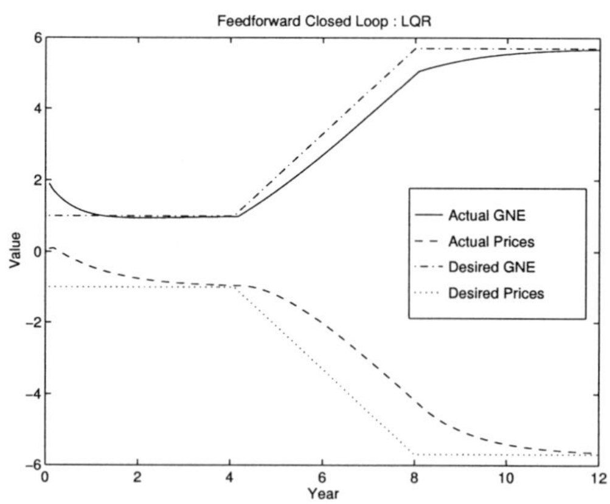

Figure 4.9: Continuous Model: Feedforward Method Using LQR

Eigenvalues of $(A - BK)$			
Continuous		Discrete	
Pole Placement	LQR	Pole Placement	LQR
−5.0000	−11.8173	0.2500	0.0012
−4.1667	−9.9484	0.3250	0.3964
−3.3333	−1.4235	0.4000	0.9017
−2.5000	−0.8122		

Table 4.3: Feedforward Method: Eigenvalues of Closed Loop State Equation Driving Matrix

Figure 4.3 (on page 124). The tracking error for the Feedforward method using the pole placement technique is 0.08, and for the linear quadratic regulator is 0.15 compared to 0.04 and 0.08, respectively, for the Integral method.

It is interesting to note the effects of perturbations on the growth vector (γ). Perturbing this vector by as much as 100% does not make a substantial difference on the overall results. In this case the tracking errors are 1.8% and 3.8% different from that without the perturbation, for the pole placement technique and the linear quadratic technique respectively.

For both the pole placement and linear quadratic regulator techniques, the transfer system matrix is reasonably well-conditioned, with $\kappa_2(G_{pp})$ of 3.2, and $\kappa_2(G_{lq})$ of 1.91. Small parameter perturbations to the system matrices will not significantly effect the results.

The third illustration is the same discrete example used to illustrate the Integral method in Section 4.3.4 (on page 129). Again pole placement and linear quadratic regulator control techniques are used to compute the state feedback gain matrix. In the latter case the weights used are, again, unity for the controls; and, for the states, 100 for each of the output states and 10 for the other state. The resultant eigenvalues of the closed loop state equation driving matrix are given in Table 4.3. The Table also contains the eigenvalues when the pole placement technique is used. Again, these parameters are chosen to aid comparison between the methods.

The fixed point integration technique was used only for a ramp input, but the model was

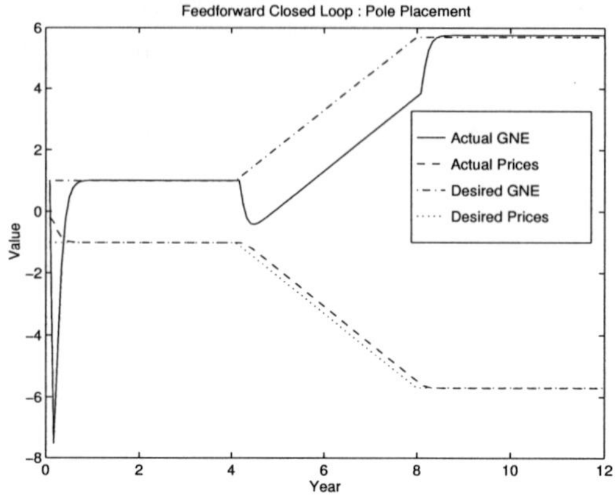

Figure 4.10: Discrete Model: Feedforward Method Using Pole Placement

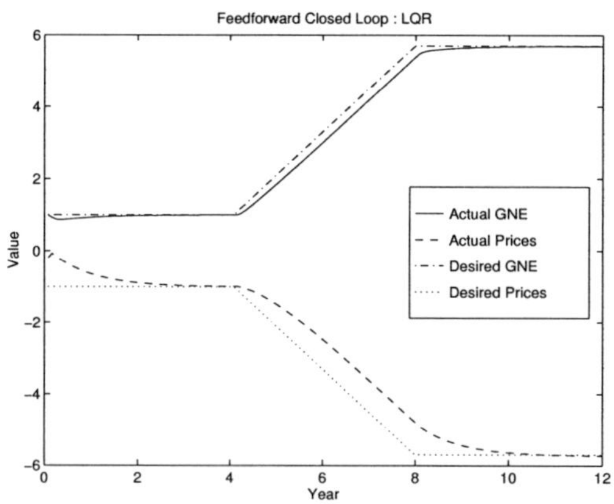

Figure 4.11: Discrete Model: Feedforward Method Using LQR

run out against the combined step and ramp trajectories as has been used in the previous examples. The results are presented in Figures 4.10 and 4.11; and compare with Figures 4.5 and 4.6 (on pages 130 and 131, respectively). The Figures show a poorer performance than the Integral method; and a particularly poor performance for the GNE output. The tracking error for this pole placement is 0.25. For the linear quadratic regulator control technique, the tracking error is 0.09. Again, perturbing the growth vector (ω) does not substantially effect the results with a doubling of the vector resulting in a 4% change and a 3.4% change for the pole placement and linear quadratic technique tracking errors respectively.

While the Feedforward method did perform well with the first illustration (Figure 4.7), it does not perform as well as the Integral method in both these illustrations. While the method tracks the desired output trajectories better than a simple closed loop system, it does not track the trajectories well. This is particularly the case in the discrete illustration using the pole placement technique (Figure 4.10).

Unlike the Integral method, the Feedforward method is not continually adjusting to the error between desired and actual output trajectories. It depends on correct model inversion to *ex ante* determine the correct reference. The method is more subject to numeric problems. The discrete illustration using the pole placement technique (Figure 4.10) results in a transfer system matrix that is not well-conditioned with $\kappa_2(G_{pp})$ of 320. The closed loop state equation driving matrix is also not well-conditioned with $\kappa_2(A - BK)$ of 1128 for K chosen by pole placement. These condition numbers indicate that any numeric problems are substantially magnified. This is the probable source of the poor performance.

Another source of error may be due to the fact that the output trajectories are an initial step followed by a combination of plateaus and ramps. This combination with the models may help add to the reduced performance. For the discrete time illustration using pole placement, with desired outputs to be only a set of ramps but with all other things the same, then the system achieves the desired output. The tracking error is reduced to 0.09; and, again, it is the GNE output that performs the worst.

This illustration shows the importance of numeric issues, and also how the condition number of the transfer system matrix can be used as a metric for these issues.

4.5 Linear Quadratic Output Tracking

4.5.1 Methodology

The final form for tracking a set of ramps to be considered is the Linear Quadratic Output Tracking method. This is a well known method (Anderson and Moore, 1990; Bryson and Ho, 1975) and is not restricted to tracking ramps. The continuous time case was considered in Chapter 2 (particularly in Section 2.7 on page 60), so the emphasis here is on the methodology using the discrete time case.

The Linear Quadratic Output Tracking method, as with the Feedforward method, requires a prior knowledge of the desired output trajectory. It does have the significant advantage, though, that it uses this prior knowledge to *predict* necessary changes in control. It is a pro-active method. This compares with the Feedforward and Integral methods which respond *ex post* to the changes in desired output. The Linear Quadratic method also uses prior knowledge of the desired controls.

The Linear Quadratic method defines a performance index, J, which is based on the difference between the actual and desired outputs and controls. This performance index is minimised subject to the linear system (Equations 4.6 and 4.7). Specifically, define the discrete time quadratic performance index as

$$J = \frac{1}{2}(\bar{\mathbf{y}}_{t_f}^T Q_{t_f} \bar{\mathbf{y}}_{t_f}) + \frac{1}{2}\sum_{t=1}^{t_f-1}(\bar{\mathbf{y}}_t^T Q_t \bar{\mathbf{y}}_t + \bar{\mathbf{u}}_t^T R_t \bar{\mathbf{u}}_t) \tag{4.40}$$

where $\bar{\mathbf{y}}$ is the deviation between actual and desired output so that $\bar{\mathbf{y}}_t = \mathbf{y}_t - \mathbf{y}_{d,t}$. Likewise $\bar{\mathbf{u}}$ is the deviation between actual and desired control vector. The time horizon is from time period 1 to t_f. The weighting matrices Q, for output, and R, for controls, can vary over the time horizon; so that, for example, future values may be discounted. The first term in the index may allow for achieving a particular desired final output. Other forms of the performance index, such as those with time-invariant, or cross-product weights, are often also used. This discrete case is analogous to the continuous time case considered in Chapter 2.

The solution to this constrained optimisation problem is given in the Appendix to this Chapter (in Section 4.7. It results in a control rule of

$$\mathbf{u}_t = -R_t^{-1} B^T L_t \{\mathbf{g}_{t+1} + S_{t+1}[A\mathbf{x}_t + B\mathbf{u}_{d,t} + E]\} + \mathbf{u}_{d,t} \tag{4.41}$$

where S_t is given by the Riccati equation

$$S_t = N_t L_t S_{t+1} + C^T Q_t C, \qquad S_{t_f} = C^T Q_{t_f} C \tag{4.42}$$

and $\mathbf{g}_t$ by the predictor equation

$$\mathbf{g}_t = N_t L_t \{\mathbf{g}_{t+1} + S_{t+1}[B\mathbf{u}_{d,t} + E]\} + C^T Q_t D\mathbf{u}_{d,t}, \qquad \mathbf{g}_{t_f} = Q_{t_f} C^T \mathbf{y}_{d,t_f} \tag{4.43}$$

with

$$L_t = [I + S_{t+1} B R_t^{-1} B^T]^{-1} \qquad N_t = A^T - C^T Q_t D R_t^{-1} B^T \tag{4.44}$$

Thus this technique requires the solution of the Riccati and predictor equations (backwards in time) before the solution of the closed loop model (forward in time).

The control rule (Equation 4.41) can be written as

$$\mathbf{u}_t = \eta_t - K_t \mathbf{x}_t \tag{4.45}$$

where

$$\begin{aligned} K_t &= R_t^{-1} B^T L_t S_{t+1} A \\ \eta_t &= \mathbf{u}_{d,t} - R_t^{-1} B^T L_t \{\mathbf{g}_{t+1} + S_{t+1}[B\mathbf{u}_{d,t} + E]\} \end{aligned} \tag{4.46}$$

which is a time-varying state feedback gain. Though, if the time horizon spans to the steady state ($t_f \to \infty$) and the weighting matrices (Q and R) are time-invariant, then the feedback gain (K) will be time-invariant. In this case Equation 4.45 is of the form as used in both the Integral and Feedforward methods (that is Equation 4.8).

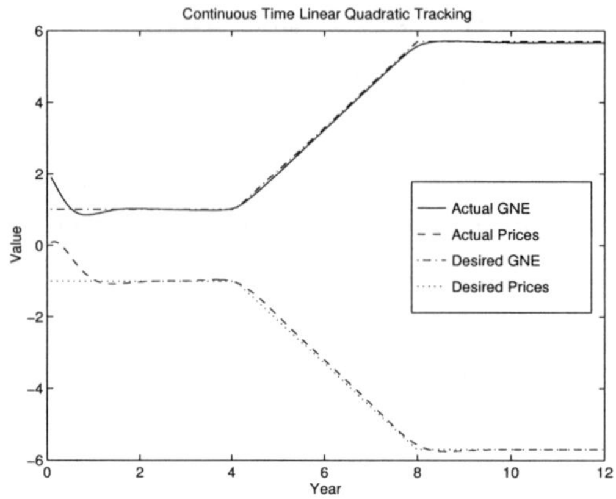

Figure 4.12: Continuous Model: Linear Quadratic Tracking

A set of metrics for the effects of parameter perturbation can also be obtained for the Linear Quadratic Tracking method. As can be seen from the Appendix, this method results in a constrained optimisation problem which can be written as an expanded system of difference equations. This is known as the Hamiltonian Equation (Equation 4.62 on page 150). The driving matrix of this equation, the Hamiltonian matrix is given by

$$H_t = \begin{bmatrix} A + BR_t^{-1}B^T N_t^{-1}C^T Q_t & -BR_t^{-1}B^T N_t^{-1} \\ -N_t^{-1}C^T Q_t C & N_t^{-1} \end{bmatrix} \tag{4.47}$$

Again, the condition of this matrix and, more importantly, the condition of the matrix formed as rows from its eigenvectors are useful metrics for the effects of parameter perturbation on this method.

4.5.2 Examples

The Linear Quadratic Output Tracking method is, again, illustrated in both the continuous and discrete time cases. The results for the continuous time case are presented visually in Figure 4.12, and Figure 4.13 presents the discrete time case. The same models and similar scenarios have been used as in the illustrations of the previous methods.

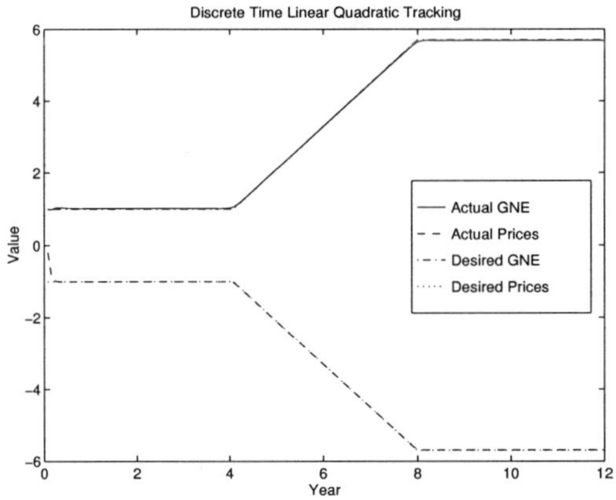

Figure 4.13: Discrete Model: Linear Quadratic Tracking

The simulation for the continuous time case is: first the same desired output trajectory is used; second, the desired control is set to zero ($\mathbf{u}_{d,t} = \mathbf{0}$) for the entire time horizon; and third, time-invariant weights are used with the control weighted at unity (R_t is the 2×2 identity matrix), and GDP weighted at 325 and the price level at 1250 (giving Q_t of the 2×2 diagonal matrix with 325 and 1250 as its diagonal elements). This results in a scenario that is comparable with the illustrations of the Feedforward and Integral Methods. As the weights are constant, so is the feedback gain matrix. The eigenvalues of the closed loop state equation driving matrix are given in Table 4.4.

It can be seen from Figure 4.12 that the method results in a controller that tracks the desired output trajectories very well. The tracking error is 0.004, which is better than in the other methods for this model.

One other point can be illustrated with this example. Using the same scenario except with increasing the output weights, results in the Riccati and predictor equations (Equations 2.80 and 2.81, on page 62) becoming unstable, so that the method cannot be used. The combination of the truncation error associated with the numerical solution technique for these differential equations, and the fact that the state equation driving matrix is ill-

Eigenvalues of $(A - BK)$	
Continuous	Discrete
$-2.7772 \pm 1.2185\imath$	0.3513
$-2.0938 \pm 2.0743\imath$	0.1528
	0.5203

Table 4.4: LQT Method: Eigenvalues of Closed Loop State Equation Driving Matrix

conditioned are the probable cause. This indicates a limitation of the method. Note that, as the state equation driving matrix is ill-conditioned, so, too, will be the Hamiltonian matrix (in Equation 2.75). The matrix formed from the eigenvectors of the Hamiltonian is also poorly conditioned, with a condition number of 1356.

Figure 4.13 presents the outputs of the discrete model when used with the Linear Quadratic Output Tracking method. The same desired trajectory for outputs was used as in the previous examples. The desired control is zero, so that $\mathbf{u}_{d,t} = \mathbf{0}$, $\forall t$. Time-invariant weights were also used with the controls being weighted at unity and the outputs each being weighted at 1000, so that R_t is the 2×2 identity matrix and Q_t is the diagonal matrix with 1000 on each diagonal element for the entire time horizon. Again, the values are chosen to be similar to other scenarios. As the weights are time-invariant, so too will be the feedback gain matrix. The eigenvalues of the resultant closed loop state equation driving matrix are given in Table 4.4.

This discrete model tracks the desired output well, as can be seen from Figure 4.13, with a tracking error of 0.005. This is the best of all illustrations. This and the continuous illustration show the superior output tracking performance of the method.

The discrete model also results in a poorly conditioned Hamiltonian matrix (Equation 4.47), which has a condition number of 1288392. The condition of the matrix formed by its eigenvectors is also poorly conditioned with a condition number of 1917. These show that the method is sensitive to small parameter perturbations.

4.6 Conclusion

This Chapter has considered three methods of tracking a set of ramps (and steps) trajectories for multiple input multiple output systems. The three methods were the Integral method, which integrates out the error; the Feedforward method, which inverts the model to determine a feedforward gain; and, the standard Linear Quadratic Output Tracking method, which determines a control rule to minimise the distance between the actual output and the desired output trajectories. Both the Feedforward and Linear Quadratic Output Tracking methods require prior knowledge of the desired outputs. The Linear Quadratic Output Tracking method has the advantage of being more general to the class of desired outputs. It is, though, the most mathematically complex of the methods and thus can be more prone to misspecification and uncertainty errors (Doyle et al., 1992).

The methods were developed and then illustrated on low order economic models. Continuous time and discrete time models were considered. The illustrations were designed to point to the advantages and disadvantages of the methods as far as the tracking of a set of desired outputs is concerned.

In summary, with the illustrations here, the Integral method using state augmentation has the advantage of good tracking performance, but may result in the loss of controllability and observability. The Feedforward method can, too, have good tracking performance, but it may be more susceptible to numeric problems than the Integral method. The Linear Quadratic Output Tracking method has superior tracking performance, though it, too, may be susceptible to numeric problems. This method, as with the Feedforward method, requires a prior knowledge of the desired output trajectory. But it uses this knowledge to predict future control changes. It is a pro-active control strategy rather than a reactive strategy as are the other two methods.

This Chapter has been concerned with tracking sets of ramps for multiple input multiple output models. The emphasis has been on the ability of the control strategy to attain the desired output trajectory. Ramps have been the concern as these fit naturally with conventional economic models, and the objective is to use control strategies for policy analysis with such models.

This Chapter has extended the servo tracking problem to multiple input, multiple output, ramps. The Chapter has also shown how the condition number of the transfer function matrix can be a useful metric in judging the performance of the controller.

4.7 Appendix: Discrete Linear Quadratic Output Tracking

This Appendix derives the discrete time Linear Quadratic Output Tracking problem of the type used in this Chapter. For later use in this book, a more general form will be considered in that the model is not necessarily square, and has time varying parameters. The continuous time problem was consider in Chapter 2 (Section 2.7 on page 60). This Appendix follows the same approach to the problem.

The problem will be formulated as: minimise the objective function, given by

$$J = \frac{1}{2}(\bar{\mathbf{y}}_{t_f}^T Q_{t_f} \bar{\mathbf{y}}_{t_f}) + \frac{1}{2}\sum_{t=1}^{t_f-1}(\bar{\mathbf{y}}_t^T Q_t \bar{\mathbf{y}}_t + \bar{\mathbf{u}}_t^T R_t \bar{\mathbf{u}}_t) \tag{4.48}$$

where the bars indicated deviations from desired trajectories so

$$\bar{\mathbf{y}}_t = \mathbf{y}_t - \check{\mathbf{y}}_t \qquad \bar{\mathbf{u}}_t = \mathbf{u}(t) - \check{\mathbf{u}}_t \tag{4.49}$$

and $\check{\mathbf{y}}_t$ and $\check{\mathbf{u}}_t$ are the desired outputs and controls and are given $\forall t$.

The objective function is minimised subject to the model, given by

$$\mathbf{x}_{t+1} = A_t\mathbf{x}_t + B_t\mathbf{u}_t + E_t \tag{4.50}$$

$$\mathbf{y}_t = C_t\mathbf{x}_t + D_t\mathbf{u}_t \tag{4.51}$$

with $\mathbf{x}_0$ given.

For the analysis here, $\mathbf{x}_t, E \in \Re^n$; $\mathbf{y}_t, \check{\mathbf{y}}_t \in \Re^m$; $\mathbf{u}_t, \check{\mathbf{u}}_t \in \Re^p$; $A_t \in \Re^{n\times n}$; $B_t \in \Re^{n\times p}$; $C_t \in \Re^{m\times n}$; $D_t \in \Re^{m\times p}$; $Q_t \in \Re^{m\times m}$ and is positive definite; and, $R_t \in \Re^{p\times p}$ and is positive definite. The problem can be solved with less restrictive conditions, but these are adequate for this book.

Define the Hamiltonian sequence

$$\mathcal{H}_t = \frac{1}{2}(\bar{\mathbf{y}}_t^T Q_t \bar{\mathbf{y}}_t + \bar{\mathbf{u}}_t^T R_t \bar{\mathbf{u}}_t) + \lambda_{t+1}^T (A_t \mathbf{x}_t + B_t \mathbf{u}_t + E_t) \tag{4.52}$$

Now this constrained optimisation must satisfy the following three conditions.

Firstly,

$$\frac{\partial \mathcal{H}_t}{\partial \mathbf{u}_t} = \mathbf{0} = \bar{\mathbf{u}}^T R_t + \lambda_{t+1}^T B_t \tag{4.53}$$

which gives the optimal control

$$\mathbf{u}_t = -R_t^{-1} B_t^T \lambda_{t+1} + \check{\mathbf{u}}_t \tag{4.54}$$

Secondly,

$$\frac{\partial \mathcal{H}_t}{\partial \lambda_{\ t}} = \mathbf{0} = A_t \mathbf{x}_t + B_t \mathbf{u}_t + E_t - \mathbf{x}_{t+1} \tag{4.55}$$

which gives the state equation

$$\begin{aligned} \mathbf{x}_{t+1} &= A_t \mathbf{x}_t + B_t \mathbf{u}_t + E_t \\ &= A_t \mathbf{x}_t - B_t R_t^{-1} B_t^T \lambda_{t+1} + B_t \check{\mathbf{u}}_t \end{aligned} \tag{4.56}$$

with the boundary condition

$$\mathbf{x}_0 = \zeta \tag{4.57}$$

Thirdly,

$$\frac{\partial \mathcal{H}_t}{\partial \mathbf{x}} = \mathbf{0} = A_t^T \lambda_{t+1} + C_t^T Q_t \bar{\mathbf{y}}_t - \lambda_t \tag{4.58}$$

which gives the co-state equation

$$\begin{aligned} \lambda_t &= A_t^T \lambda_{t+1} + C_t^T Q_t \bar{\mathbf{y}}_t \\ &= A_t^T \lambda_{t+1} + C_t^T Q_t C_t \mathbf{x}_t + C_t^T Q_t D_t \mathbf{u}_t + C_t^T Q_t \check{\mathbf{y}}_t \\ &= N_t \lambda_{t+1} + C_t^T Q_t C_t \mathbf{x}_t + C_t^T Q_t \check{\mathbf{y}}_t \end{aligned} \tag{4.59}$$

where

$$N_t = A_t^T - C_t^T Q_t D_t R_t^{-1} B_t^T \tag{4.60}$$

and, with the boundary condition

$$\lambda_{t_f} = C_{t_f}^T Q_{t_f} \check{\mathbf{y}}_{t_f} \tag{4.61}$$

This produces a coupled set of linear differential equations which form a two point boundary value problem with mixed boundary conditions.

While it is not directly required for this derivation, the system can be written as the Hamiltonian Equation of

$$\begin{aligned} \begin{bmatrix} \mathbf{x}_{t+1} \\ \lambda_{t+1} \end{bmatrix} &= \begin{bmatrix} A_t + B_t R_t^{-1} B_t^T N_t^{-1} C_t^T Q_t C_t & -B_t R_t^{-1} B_t^T N_t^{-1} \\ -N_t^{-1} C_t^T Q_t C_t & N_t^{-1} \end{bmatrix} \begin{bmatrix} \mathbf{x}_t \\ \lambda_t \end{bmatrix} \\ &+ \begin{bmatrix} N_t^{-1} C_t^T Q_t & N_t^{-1} C_t^T Q_t D_t \\ B_t R_t^{-1} B_t^T N_t^{-1} C^T Q & B_t R_t^{-1} B_t^T N_t^{-1} C^T Q D_t + B_t \end{bmatrix} \begin{bmatrix} \check{\mathbf{u}}_t \\ \check{\mathbf{y}}_t \end{bmatrix} + \begin{bmatrix} E_t \\ O \end{bmatrix} \end{aligned} \tag{4.62}$$

with O being the n element zero vector.

Specifically, it is the Hamiltonian Equation driving matrix which plays the important role. The Hamiltonian Equation driving matrix is referred to as H_t in this Chapter.

The problem can be solved analytically by the sweep method. Try for a solution. Let

$$\lambda_t = S_t \mathbf{x}_t + \mathbf{g}_t \tag{4.63}$$

then

$$\lambda_{t+1} = S_{t+1} \mathbf{x}_{t+1} + \mathbf{g}_{t+1} \tag{4.64}$$

Substitute the state equation (Equation 4.56) giving

$$\lambda_{t+1} = L_t[\mathbf{g}_{t+1} + S_{t+1}(A_t \mathbf{x}_t + B_t \check{\mathbf{u}}_t + E_t)] \tag{4.65}$$

with

$$L_t = [I + S_{t+1} B_t R_t^{-1} B_t^T]^{-1} \tag{4.66}$$

and I being the $n \times n$ identity matrix.

Using the co-state equation, Equation 4.59, and substituting for λ_t from Equation 4.63 and rearranging gives

$$N_t \lambda_{t+1} = [S_t - C_t^T Q_t C_t]\mathbf{x}_t + \mathbf{g}_t - C_t^T Q_t D_t \check{\mathbf{u}}_t + C_t^T Q_t \check{\mathbf{y}}_t \tag{4.67}$$

Substituting Equation 4.65 into Equation 4.67, rearranging into the form of Equation 4.63; and, then, by equating coefficients, the Riccati Equation

$$S_t = N_t L_t S_{t+1} + C_t^T Q_t C_t \tag{4.68}$$

and the predictor equation

$$\mathbf{g}_t = N_t L_t \{\mathbf{g}_{t+1} + S_{t+1}[B_t \check{\mathbf{u}}_t + E_t]\} + C_t^T Q_t D_t \check{\mathbf{u}}_t \tag{4.69}$$

are obtained.

The boundary conditions are also obtained by substitution and equating coefficients with Equation 4.63. This results in

$$S_{t_f} = C_{t_f}^T Q_{t_f} C_{t_f} \qquad \mathbf{g}_{t_f} = Q_{t_f} C_{t_f}^T \check{\mathbf{y}}_{t_f} \tag{4.70}$$

The optimal control rule can then be rearranged so that it is not necessary to calculate the co-states. This rule is then given by

$$\mathbf{u}_t = -R_t^{-1} B_t^T L_t \{\mathbf{g}_{t+1} + S_{t+1}[A_t \mathbf{x}_t + B_t \check{\mathbf{u}}t + E_t]\} + \check{\mathbf{u}}_t \tag{4.71}$$

Chapter 5

VISUALISATION TECHNIQUES

5.1 Introduction

One of the aims of this work is to develop tools so that the ideas presented in it can be explained to, and (hopefully) used by, economists. Many economists, especially practising (as distinct from academic) economists, do not have the time to understand the concepts and mathematics presented in this book. An aim is, thus, to develop and demonstrate software tools so that these ideas can be more easily accessed.

It is not surprising that many of the concepts and approaches of control theory fit comfortably with engineers, for this is where it has its roots. Some of these, such as constrained optimisation, also fit comfortably with the typical approaches used in economics; but others, like root locus, do not fit so comfortably. It is an aim of this book to show how some of the approaches and tools developed in the control community can be useful in the analysis of dynamic economic systems. This is not only for the more abstract ideas, but also for many of the simple practical ideas.

This Chapter is concerned with the use of visualisation in the simulation and analysis of dynamic economic models. In so doing, it presents the schematic approach which is common for control engineers, but rarely used by economists. It is argued that this approach and its resultant simulation software is useful when considering dynamic economic models, and further, is a means by which the ideas of this book can be disseminated to economists.

The Chapter has two primary aims. First there is the visual presentation and simulation of models and algorithms from earlier Chapters, leading to the development of a simulation environment that is used in later Chapters. The second aim is to develop a simulation and data visualisation environment where policy implications can be easily seen. This enables trade-offs and policy effects to be examined. For example, frequently with control type exercises in economics fixed weight objective functions and specific state targets are used in an optimal control setting. An example is the work of Murphy and Brennan (1993). But the really interesting questions concern what are the effects of changing the output and control weightings to examine the trade-offs between particular policy strategies. Questions such as: 'what are the effects of more emphasis on monetary rather than fiscal policy to achieve the same objective', show these trade-offs. Further, temporal targets are of more interest, and lead to questions such as: 'is it possible to design policy to slowly reduce unemployment and still maintain an inflation target'. There are also the questions of robustness of the policy to exogenous shocks, like a rapid change in the Rupiah to Australian dollar exchange rate, on the solution. An aim of this Chapter is to build a simulation environment where such questions can be simulated with models by practising economists.

Simulation exercises for economic systems can require substantial overheads in the development of the simulation system, and can produce large quantities of output. Visualisation techniques can be used to help in both these areas. Firstly by using visual simulation system's development tools which make the process of simulation easier. Secondly by using data visualisation tools which make it easier to comprehend large quantities of data. These two areas are the focus of this Chapter.

The next section (Section 5.2) considers the use of visual simulation tools through simulation of the Currie and Levine (1993a) model examined in Chapter 2 (Section 2.5 on page 36). This is a case study approach of the type frequently used in visualisation (Robertson and Silver, 1995). Consideration of the use of visual simulation system's development tools, is by gradually increasing the complexity of the simulation through three illustrations: firstly simulation the model itself in open loop; then the system is extended to learn the forward looking variables; and, finally, optimal learning and optimal control are illustrated. The illustrations become increasingly more mathematically complex, and the aim is to examine the use of visual simulation development tools to gradually build a sophisticated simulation.

A number of commercial visual simulation system's development tools which are suitable for use with economic models are available. These include BuildSim by Tangent Systems, VisSim by Visual Solutions, ModSim by CACI Products and Simulink by The Math Works. Most of these run on a variety of computing operating systems including Microsoft Windows and Unix (X-Windows) giving a seamless transfer to differing degrees of computing power. Some have, or are coupled with, data visualisation tools. The simulation software tool considered here is Simulink (Mathworks, 1994*b*) which is designed to simulate dynamic systems, and is commonly used in the control community. Objects (or blocks) are visually selected and combined using a 'drag and drop' approach to build simulation systems. The objects may be, for example, components of equations, equations or output generators. Systems of objects can be grouped into higher order objects. This is particularly useful for 'information hiding' so that the complexity of the system can be more readily comprehended. Parameters, such as covariances, can be interactively altered during a simulation so that the implications can be easily examined. During a simulation data can be viewed graphically or stored for separate analysis or visualisation.

Section 5.3 of this Chapter uses this visual simulation environment to present some of the techniques developed in Chapter 4. It also extends the Chapter 4 results by the inclusion of an observer, and introduces integral control without state augmentation. The Section shows, for example, how the trade-offs between the control policies necessary to track ramp outputs can be easily compared; and, how the addition of the observer is a visually simple extension.

Section 5.4 looks at numerical issues and show how these can easily be considered in such an environment. Usually with the building of economic models, the focus is on model specification and upon the statistical properties of the model. Rarely is much emphasis given to the numerical solution technique, or properties such as inter sample behaviour. This section considers how the numerical solution technique can have significant effects on the simulation output. It shows how a visual simulation environment allows the easy examination of this type of issue.

Simulations can produce large quantities of output data, and it is in the comprehension of this data that the tools for data visualisation can be useful. This is considered in Section 5.5, and is illustrated by considering a reasonably simple simulation question of how multipliers produced by a model vary with differing model parameters.

5.2 Visual Simulation Systems Development

This Section considers some simulations of the Currie and Levine (1993*a*) model using the simulations system's development tool. A relatively simple simulation is used firstly, and gradually the mathematical sophistication of the simulation is developed. The aim is to consider the simulations development and comprehension under the development tool.

SIMULINK, the development tool, is designed to simulate dynamic systems. Its roots are in control theory, and it has not been used to any great extent with economic models. The models it simulates can be expressed in continuous time (as is the case here), discrete time (as in Section 5.3), or a combination of both. Most dynamic economic models are expressed in discrete time. SIMULINK uses objects in a windowing operating system; with the simulation system using one or more windows. Within a simulation window, objects can be copied and pasted and, similarly, they can be moved or copied between simulation windows. Simulation systems are developed by joining objects via data flow lines. This comes from the schematic block diagrams frequently used by the control community. Objects receive data via these lines and manipulate it according to their attributes. Objects are visually moved around and between windows through the use of a pointing device (e.g. a mouse), and data flows are drawn via the device.

Some objects, such as those for graphically displaying data, have their own windows. Higher level objects, which are made up of groups of lower level objects, also have their own window. These windows can be opened and closed during the simulation so that the user can decide upon the level of detail of the simulation system or output data they wish to see.

The simulation is controlled through a series of menus on a simulation environment window.

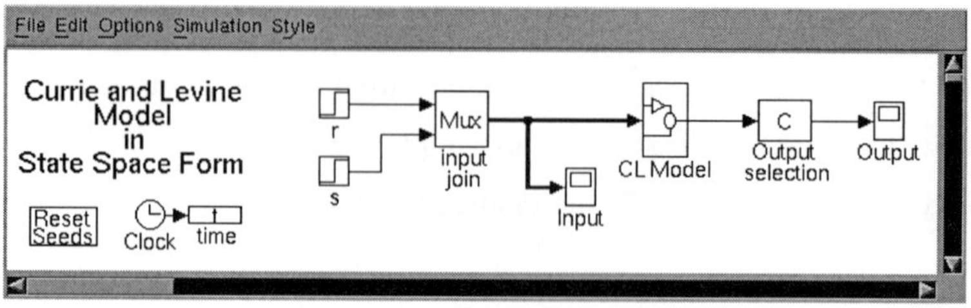

Figure 5.1: Main Simulation Window for Model in State Space Form

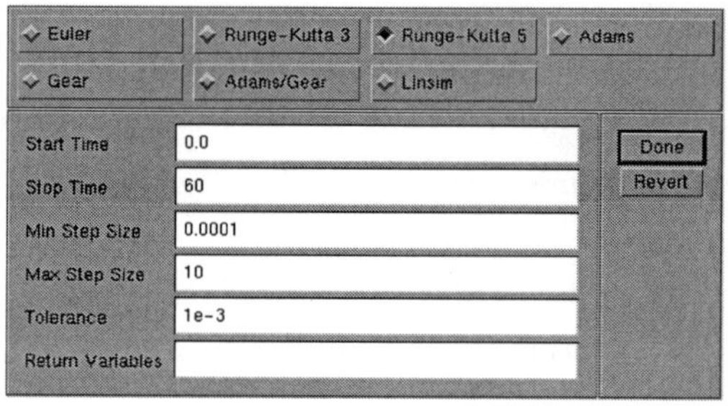

Figure 5.2: Parameters Dialogue Box Window from The Simulation Menu

Such features as the length of time to run and the numerical method used can be adjusted through these menus. Normal features such as simulation environment saving, font and colour choices are also available through these menus.

An example of a main simulation window is given in Figure 5.1. This is the main simulation window for simulating the Currie and Levine (1993*a*) model in state space form and is considered in more detail below. The window shows the menu options that are available to the user when performing the simulation. An example of one of the options from the Simulation menu is shown in Figure 5.2; and this dialogue box allows for the selection of the detailed parameters of the simulation, including the numerical techniques used in the simulation.

To show the seamless across platform capabilities of this visual simulation development tool,

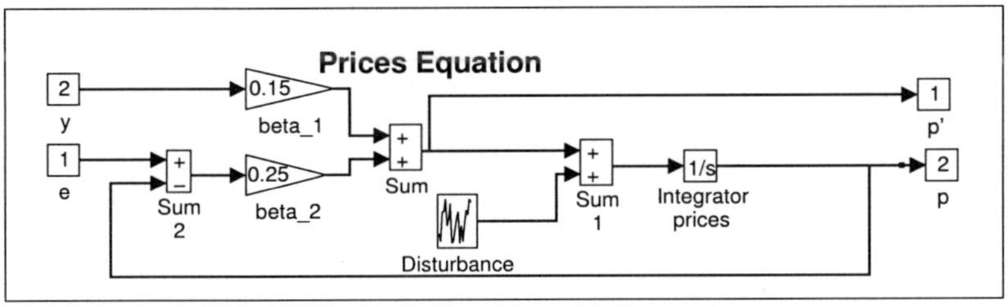

Figure 5.3: Prices Equation in Structural Form

Figures 5.1 and 5.2 were produced under the X-Windows (Unix) environment, whereas many other Figures were produced under Microsoft Windows on a PC.

The following illustrations show the use of this tool for the development of a simulation. The Figures show the contents of particular windows from simulations; that is the models themselves, rather than the actual window as in Figure 5.1.

5.2.1 The Price Equation in Structural Form

Figure 5.3 visualises the prices equation, Equation 2.50 (on page 36) in structural form. The objects that make up the equation are connected to show the data flow through the equation. This equation is made up of low level objects such as constant multiplication, summing and the generation of stochastic disturbances. Numerical integration is undertaken by the *Integrator Prices* object, and it is represented using the Laplace operator (s). The linkages from and to other equations can be clearly seen. It has been decided to include the expected inflation rate (under the rational expectations hypothesis) in this structural form so that both prices, p, and the expected inflation rate, $\dot{p}^e$ (and labelled p' in the Figure), are considered as being generated by the equation. The addition of the supply side shocks is modelled by Currie and Levine (1993a) as a random disturbance. The *Disturbance* object generates white gaussian noise of a given variance. Selecting this object allows for more detailed manipulation of the object. In this case SIMULINK presents a dialogue box which allows parameters of the noise to be altered.

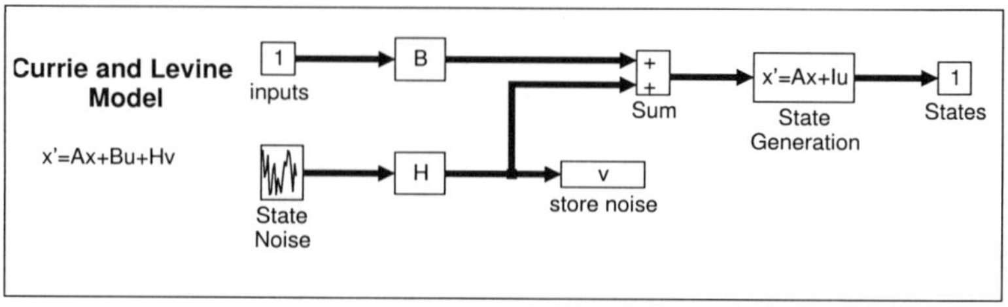

Figure 5.4: Model in State Space Form

5.2.2 The Model in State Space Form

The model in state space form is shown visually in Figure 5.4. This is the state equation, Equation 2.54 (on page 38), for the model. Here the data flow lines represent vectors rather than the scalars used in Figure 5.3, and the objects input manipulate and output vectors. One of the features of SIMULINK is that objects detect the dimensionality of their input and adjust accordingly, with objects inheriting the properties of their class. Thus the *Sum* object in Figure 5.3 could have been copied into the window given in Figure 5.4 and the object would adjust to the greater dimensionality of the input. The *Store Noise* object is a data sink for storage of data for use after the simulation.

This window also illustrates the power of the visual simulation state space structure. Simply changing the system parameter matrices (objects A, B, H, and C) allow for the simulation of a different model.

The equation itself can be considered as an object. A simulation system for the model is shown in Figure 5.1. Sources for the input data are combined and input into the *CL Model* object, and shows how information can be hidden if it is unnecessary for the current purpose. The outputs are selected through matrix multiplication, and are input into a graphical output object. Selecting the *Output* object will open its window so the model outputs can be seen. The *Reset Seeds* object resets the seeds used in the generation of the stochastic disturbances; and the *Clock* object outputs the simulation time.

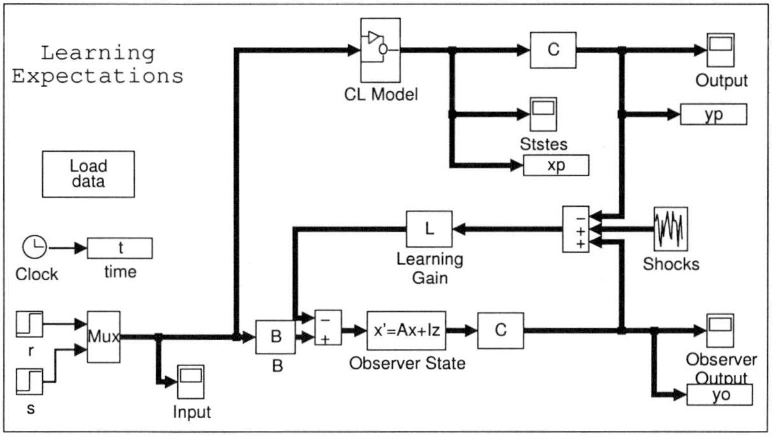

Figure 5.5: Learning Expectations

Using the simulation system presented in Figure 5.1, the model can be simulated and the results (and inputs) examined interactively. Noise variances, for example, can be easily altered and the implications seen graphically. This simulation is set up to generate maintained multipliers with a unit shock being given to interest rates (r) after a specified time period. But the effects, say, of increasing the supply-side shocks can be easily examined in this environment.

5.2.3 Learning Expectations

This illustration shows how the simulation system can be easily extended; and is done so by the inclusion of an observer to learn the forward looking expectations. The addition of the observer is shown in Figure 5.5. Here a number of graphical output objects are used to display simulation results during the simulation. The use of data sink objects for post simulation of data is also illustrated. The structure of the observer can be seen in the Figure. This simulation system also includes additional shocks (as measurement noise) so that a weaker form of the rational expectations hypothesis can be simulated as expectations 'jump' to 'news'. The *Load Data* object loads simulation parameters and other data.

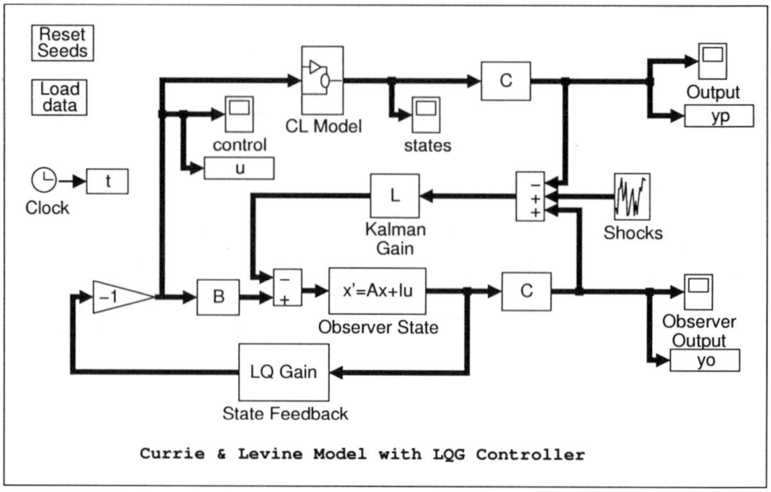

Figure 5.6: Optimal Controller and Optimal Learning

This illustration uses the basic ideas of Chapter 2 (in Section 2.4 page 26) where it was shown how an observer could be used to learn expectations variables. The mathematics is given by Equations 2.20 and 2.21 (on page 29) for the Currie and Levine (1993*c*) model, and Equation 2.22 and 2.23 (on page 30) for the observer. The learning gain is chosen by the Kalman filter technique of Section 2.4.3 (on page 35).

This illustration shows how a more complicated simulation system can be made up by the copying of other objects. All the objects used in this illustration were also used in the previous one, but some have been copied and added for this illustration. Further, while the simulation is in progress different data items can be examined by opening of the graphical output objects. The structure of the simulation is easily modified by adding more of these objects or by changing the data lines into these objects.

5.2.4 Optimal Control

Figure 5.6 shows the simulation system being expanded with the introduction of optimal control of the model. The model's inputs are now generated on the basis of the observer's states, with the feedback loop being closed with a stationary gain linear quadratic regulator. This simulation system is basically that considered in Section 2.5 (on page 36).

As can be seen from a comparison of Figure 5.5 and Figure 5.6 the expansion of the simulation environment is visually relatively simple and allows for an easier understanding of what is being undertaken by the complex mathematical simulation set up here. Optimal control generation is undertaken with the optimal learning and the resultant Linear Quadratic Gaussian problem is illustrated in Figure 5.6. This simulation is complicated mathematically with, effectively two models, two objective functions being minimised and their interactions, yet visually it is easily seen as a logical extension of the earlier illustrations.

This simulation environment allows for easy experimentation with the Currie and Levine (1993*c*) model. The learning of expectations variables can be seen visually; and, for example, its 'jumping' to 'news' can be easily modified. Further, the trade-offs between different policy settings can be visually seen with, for example, the weights between different policy settings on the linear quadratic regulator being altered during a simulation. This simulation environment makes it easy to answer questions such as: 'with other things being equal, what are the effects of the government weighting fiscal policy (s) more heavily than monetary policy (r)?'

5.3 Observers and Tracking Ramps

5.3.1 The Integral Control Method

The simulation environment can be further extended to include the algorithms of Chapter 4 for the tracking of ramps. The mathematics of that Chapter can be simpler to comprehend

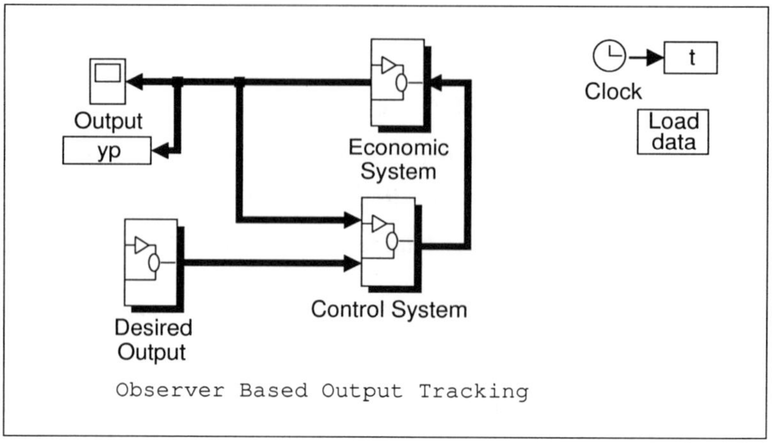

Figure 5.7: Tracking Ramps Simulation Environment

visually. Further, the algorithms can be visually easily extended to include the use of an observer. In this case, the observer based control system is separate from the economic model itself. It is this concept that is at the heart of this book: *viz.*, given some economic system, can policy analysis be undertaken upon this system via policies developed on a separate model using control techniques. Frequently the economic system is a large practical econometric model, and the control system is based upon a low order economic model. The economic system is, in this case, in line with the current econometric modelling theory, but the control system being separate, though related, can use control theory techniques.

Visually the simulation environment for the approach is given in Figure 5.7. There are two main objects in a tight loop: the *Economic System* object takes its input from the *Control System* object, which, in turn, takes as part of its inputs, the outputs from the *Economic System.* The other input into the *Control System* is the output from the exogenous *Desired Output* object. The other objects in this simulation environment have been introduced in earlier illustrations.

For this illustration, the *Economic System* is a discrete model in state space form. This state space form is the form used in this book and can be found in Equations 4.6 and 4.7

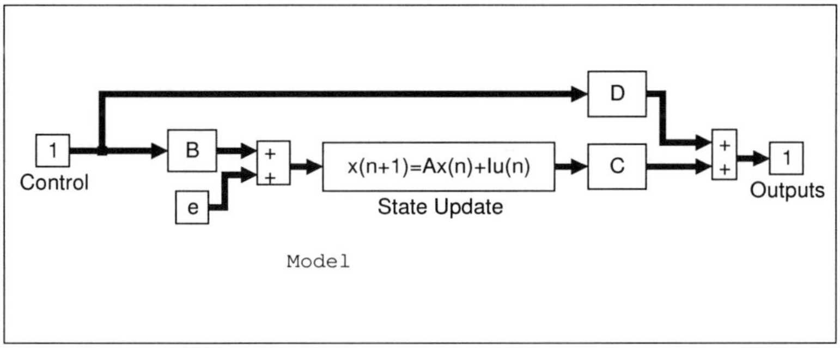

Figure 5.8: The *Economic System* Object

(on page 113). The details of this object are presented in Figure 5.8. The parameters of the model, that is objects A, B, C, D and e, are determined by the *Load Data* object in the main simulation window (Figure 5.7). The particular model in this illustration is the three state model used in Chapter 4 (in Section 3.3.5 on page 75) and developed in Chapter 3 (in Section 4.3.4 on page 129). But the important point is that any model in the state space form can be easily implemented, and, as shown in Chapter 3, this is the situation for a wide variety of economic models. This illustration also uses a discrete model to highlight the similarity of the models for visual simulation. The *State Update* object of Figure 5.8 is the discrete analogue of the *State Generation* object of Figure 5.4 (on page 158).

The *Control System* in this illustration uses the Integral Control method of Chapter 4, but is extended to an observer through output-injection from the *Economic System* object. As was considered in Section 4.3 (on page 114), the basis of the method is that model is augmented with additional states aimed at eliminating the error between desired and actual output; but, essentially, the structure of the model remains the same. This can be visually seen in Figure 5.9. The major objects in that window (e.g. *Observer Update*) have essentially the same structure as those in Figure 5.8 (e.g. *System Update*), but are the augmented system as given by Equations 4.28 and 4.29 (on page 128). The additional injection of *Desired* output into the state equation (Equation 4.28), the *Model Output* from the *Economic System* to form the observer, and the closing of the loop by the *Feedback*

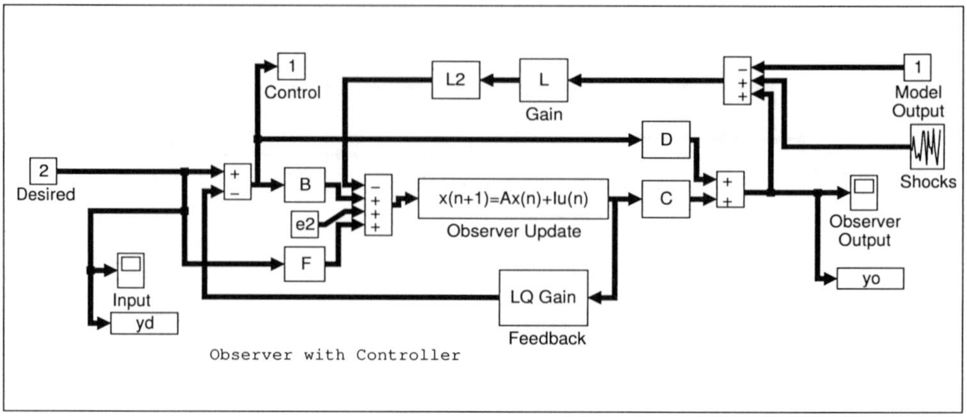

Figure 5.9: The *Control System* Object

object; can be seen as a simple visual extension of Figure 5.8, but the mathematics of this augmented system is sophisticated. Again, this highlights the power of visual simulation in explaining complex simulations. Not only is information hiding by objects being illustrated in Figure 5.7, but also in this Figure; so that complex objects appear as simple as they need be for the current situation. The aim is to present economists with powerful tools, but not to confuse issues with unnecessary details.

This particular illustration uses a linear quadratic regulator to generate the state feedback gain, and is essentially comparable to the illustration given in Figure 4.6 (on page 131). Again, selecting the *Feedback* object displays a dialogue box allowing the weights of the regulator to be chosen during a simulation. Hence policy analysis can be easily undertaken with the simulation environment.

In Chapter 4 (on page 120), it was shown that the Integral Control method used in this illustration results in the model becoming unobservable. Yet this illustration shows it being used as an observer. The augmented model becomes unobservable for it is not possible to recreate the augmenting states. But for the observer in this illustration it is only necessary to recreate the original states. Here the observer is an increased order observer as it is of

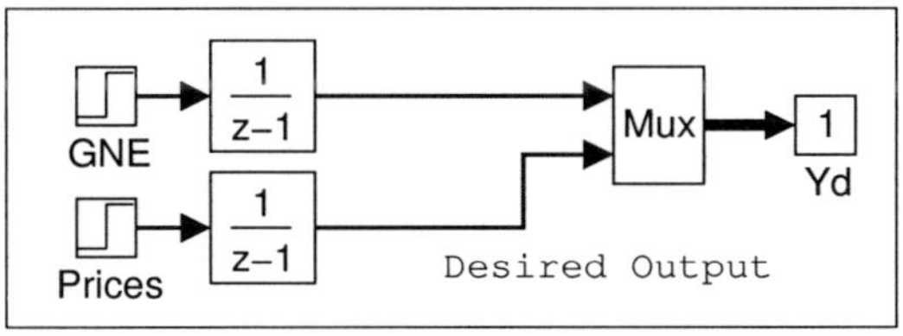

Figure 5.10: The *Desired Output* Object

higher order than the model, and, in an analogous manner to the concept of a reduced order observer (Ogata, 1990), it need only observe the non-augmenting states.

For the illustration in Figure 5.9, the *Gain* object uses pole placement to select the observer learning gain. It, too, can be altered while the simulation is running, so that the effects of altering the power of the learning component can be considered. The *L2* matrix object simply increases the dimensionality of the learning component with additional zero states. Thus *L2*, in the nomenclature of Chapter 4, is given by

$$\bar{L}_2 = \begin{bmatrix} I_{m \times m} \\ O_{2m \times m} \end{bmatrix} \tag{5.1}$$

where I is the identity matrix, O the zeros matrix, and, m the number of outputs.

Figure 5.10 presents the *Desired Output* object which is responsible for generating the desired outputs. In this illustration, a positive step in *GNE* and a negative step in *Prices* are given after an initial delay and these are summed to produce ramps of the type used in Figure 4.6 (on page 131). The summation is through the z-domain integration analogue. These objects change dynamically with the time step, which is one (Quarter) in this illustration.

This illustration has deliberately introduced a new concept into this book, in the aim of showing how visually, it is a reasonably obvious incremental step from earlier illustrations. The concept is that of a higher order observer being used to control a lower order model. Rather than it being introduced more rigoursly in Chapter 4 with the augmentation technique, it was introduced here to highlight the benefits of the visual simulation approach.

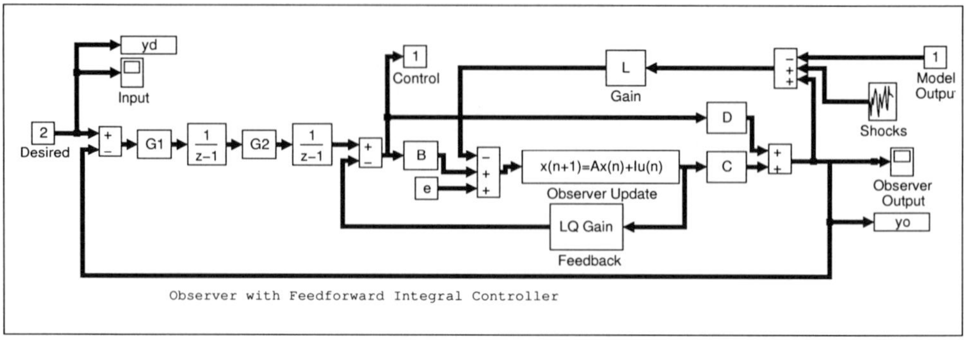

Figure 5.11: Modified *Control System* Object

5.3.2 *Integral Control Without Augmentation*

This illustration highlights some earlier points by considering a change of control methodology. In this case all objects are the same as in the previous illustration with the exception of the *Control System* object. It is modified with components simply copied from other objects. The control methodology also 'integrates out' the error between the *Observer Output* and the *Desired Output*, but does so without augmentation. Thus the observer can be of the same order as the *Economic System*. The error is 'integrated out' through the addition of an extra feedback loop and a feedforward integral components. As the desired output is a set of ramps, double integration is used. This is introduced as a series of two integrator objects with each containing an integrator for both of the two outputs. Each integrator object has its own gain. An alternative method would have been to use a state space representation (using states as in Equation 4.25 on page 127) for the double integration.

The *Control System* object's details are given in Figure 5.11. Comparing it to Figure 5.9 shows the essential differences between the two methods. The additional feedback loop and the feedforward integral component can be easily seen. Information hiding has been used so that the differences in order between this object and its counterpart in Figure 5.9 are not obvious, but the important concept of similar structure, is obvious. Visually the differences between the two methods are only simple changes, but mathematically the differences are

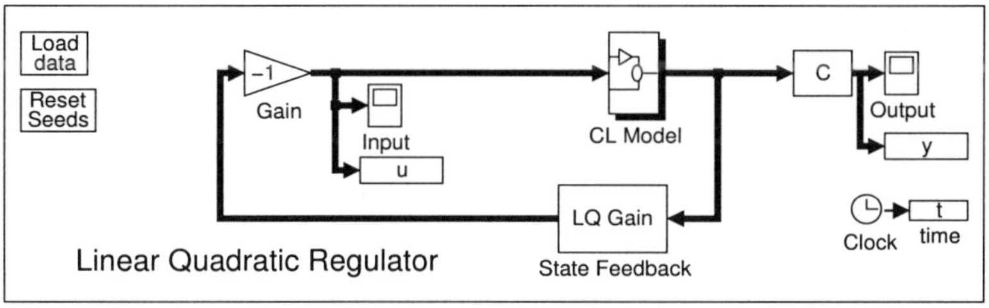

Figure 5.12: Linear Quadratic Regulator

much more complex.

One of the disadvantages of this method is that the weighting of the integral component is not incorporated into the state feedback selection procedure as it is with the augmentation method. The choice of the gains for the *G1* and *G2* objects is not as obvious in this methodology, and the selection is important as integration can create instability. On the other hand, it does not effect the controllability or observability of the system. Details of the method for a single integrator object and a desired output of a step can be found in Ogata (1995, p.461).

5.4 Numerical Issues

One of the advantages of this visual simulation environment is that it allows for the easy consideration of numerical issues. In this section, this is illustrated through the use of the Parameters Dialogue Box presented in Figure 5.2 (on page 156). For an illustration, the Currie and Levine (1993*c*) model is used in an essentially similar exercise as they undertake. Here, though, future social losses are not discounted (so that stationary gains are used), and backward looking prices expectations are assumed (so that the initial value of expected prices is given). Thus the model can be set up in the closed loop form as presented in Figure 5.12.

This illustration shows the effects of using different numerical integration routines on sim-

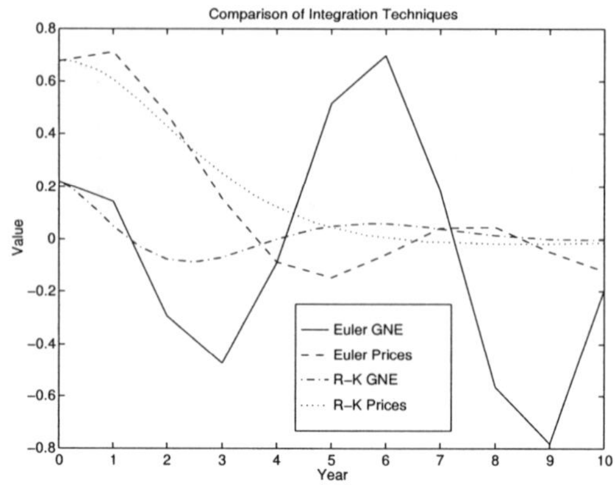

Figure 5.13: Comparison of Output from Different Solution Techniques

ulating this scenario. Currie and Levine (1993*c*) do not discuss solution methods but do state "a one year time period is assumed" (Currie and Levine, 1993*a*, p.136). So, for this illustration, a one year sample time will be used rather than the one month as in earlier illustrations. Figure 5.13 shows the outputs from this scenario with all things equal except the numerical solution technique. The Runge-Kutta technique (Burden and Faires, 1985) uses the variable step size Runge-Kutta 4/5 algorithm with the step size selected between 0.0001 and 10 such that the local truncation error is not greater than 0.001. These are the options presented in Figure 5.2. The Euler technique (Burden and Faires, 1985) uses a fixed step size of one. This is the usual technique used in economic modelling. Figure 5.13 compares the outputs under these solution techniques and shows that very different outputs are attained. The Euler solution is unstable, whereas the Runge-Kutta is not.

This illustration highlights how visual simulation allows for important issues such as numerical solution techniques to be considered, and easily presented to less numerically aware users.

5.5 Data Visualisation

This section considers the use of data visualisation tools to aid in the comprehension of output from simulations. It is easy to produce copious quantities of data from simulations such as those above; and often it can be difficult to comprehend features of the results due to the sheer volume of this data. Data visualisation tools aim to help present the data in a way to aid in its interpretation.

Here the visualisation tools of three dimensional surfaces and animation are considered. As animation cannot be illustrated in this medium, a small sample of the frames is presented. The techniques are illustrated by considering the relatively simple question of how the response of the model changes to perturbations in its parameter values. Again, the Currie and Levine (1993*c*) model is used, and the simplest simulation environment of Figure 5.1 is used to develop maintained multipliers (or the unit step response).

Consider a model with n inputs, m outputs, and that takes T time steps to arrive at its steady state or long-run solution. Let K parameters be varied, and let the ith parameter value be given by p_i. Then the number of multipliers is given by

$$M = nmT \prod_{i=1}^{K} p_i \tag{5.2}$$

So for this model with two inputs; two outputs; and, say, 120 time steps, and two parameters each given ten values; then there are 4800 multipliers. Thus even with a relatively simple simulation exercise, a large quantity of data can be produced and the basic question of how parameter perturbation effects the multipliers can be lost in the detail.

Figure 5.14 visually presents some of this data. Here the multipliers for GDP with respect to interest rates and with respect to taxes are presented as two surfaces. The multiplier is given on the vertical axis with colour used to represent value, and with time and varying parameter values for β_2 on the other axes. Viewing the surfaces from a different viewpoint, as is given in Figure 5.15, helps to show the effects of the changing parameter values. More beneficial is using animation to move the viewpoint around the surfaces. Animation can

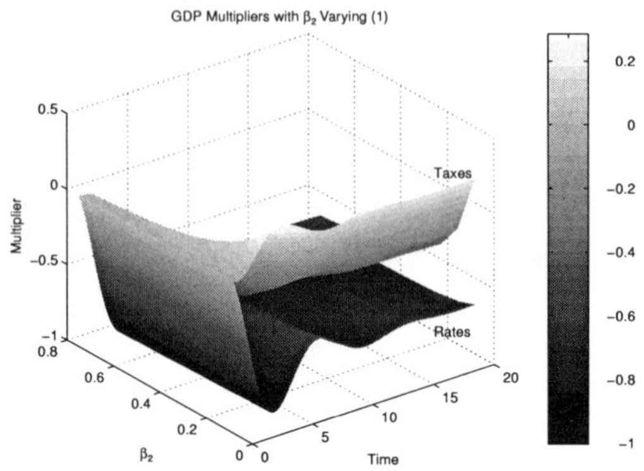

Figure 5.14: GDP Maintained Multipliers with β_2 Varying

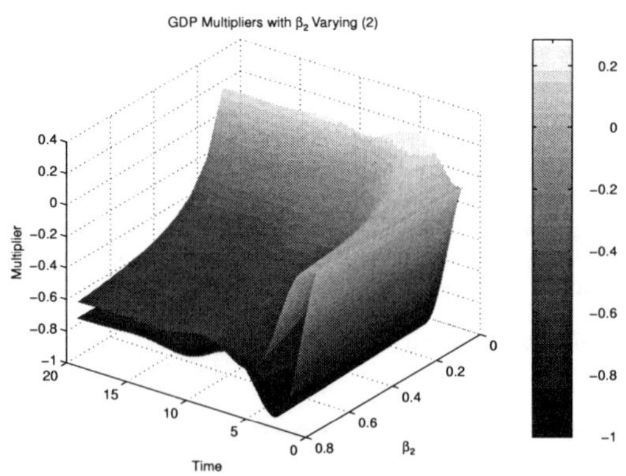

Figure 5.15: GDP Maintained Multipliers with β_2 Varying. A different view.

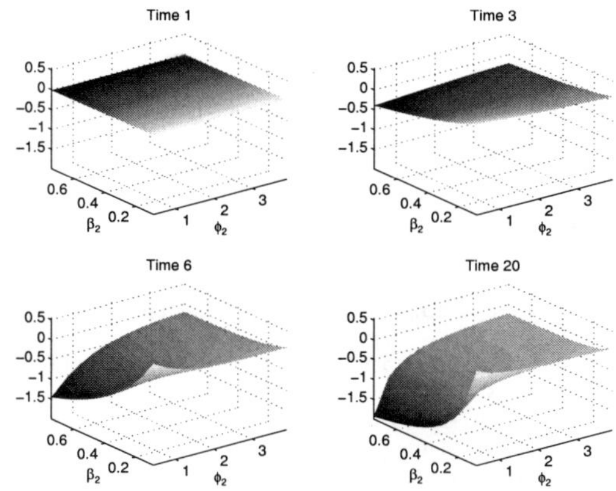

Figure 5.16: GDP Multipliers for Interest Rates with β_2 and ϕ_2 Varying.

also be useful to convey the passage of time. If two parameters are perturbed, then a surface of multipliers is produced for each time period. How this surfaces 'moves' with time shows the implications of the variation in parameters. Figure 5.16 plot the short-run and long-run multipliers of the model for GDP with respect to interest rates with two model parameters (β_2 and ϕ_2) varying. Using colour to indicate magnitude and animation to indicate time helps in the understanding of the effects of the changes in parameter values on the multipliers.

5.6 Conclusion

This Chapter has focused on visual simulation and has been concerned with two main issues. The first has been to present tools that are to be used in later Chapters of this book; and to show, visually, some of the algorithms developed earlier in this book. But of much greater importance, it has argued that visual simulation is a means by which, first, some of the complex ideas and algorithms of this book may be presented to less numerical, mathematical and computing literate users; and, second, simulations of these algorithms can be easily undertaken using visual simulations so that such users can easily investigate policy analysis questions.

The Chapter has also highlighted the benefits of data visualisation, though it is more difficult to present in this medium. Again, it is argued, that these are powerful tools that can help present ideas, concepts and results form this book.

The Chapter has shown how the visual block diagrams used by the control community are useful for analysing dynamic economic models. It has also shown the benefits of the state space representation of such models. Further, it has shown how the simulation software developed for these diagrams leads to considerable benefits when investigating dynamic economic models.

The visual simulation environment was presented by gradually increasing the complexity of simulations. The step from a mathematically simple to a mathematically complex simulation was shown to be a reasonably simple step visually. It was also shown how different types of models and different algorithms can be easily implemented in such an environment. Issues such as the solution technique for models can also be easily considered in the environment.

The main point of this Chapter is that visualisation tools are a powerful means of presenting the sophisticated algorithms used in this book.

Part II

Nonlinear Systems

Chapter 6

OPTIMAL CONTROL WITH NONLINEAR MODELS: RECENT STUDIES

6.1 Introduction

There has been a large number of studies using optimal control theory with economic models for policy analysis. This Chapter reviews a selection of recent studies which use optimal control techniques with nonlinear models. The selection presents the thrust of the current approaches, and considers studies that are particularly relevant to the book.

The early approach was to linearise the nonlinear econometric model and to use linear quadratic optimal control techniques (Chow, 1981*a*; Kendrick, 1981). Some recent studies developed from this approach are considered in Section 6.2. The first set of studies, in Section 6.2.1, focus on the uncertainty of the model; and, the second set, in Section 6.2.2, focus on the difficulties of linearising a very large nonlinear econometric model.

The second thrust of current research is into nonlinear optimisation; and, a selection of these studies is considered in Section 6.3. The first set of studies, in Section 6.3.1, focus on rational expectations models, and use the stacking over time approach to generate a single optimisation. The second set of studies, in Section 6.3.2, are, like this work, Australian studies.

A final study is considered in Section 6.4. This is an old study (undertaken in 1970s) and is included because, like this work, it uses a separate model based control approach to develop policy instruments.

6.2 Linearisation Approaches

Given a nonlinear econometric model and some objective function, one set of approaches has been to transform this original problem into a linear quadratic problem. Then, the approaches use linear quadratic optimal control techniques on the transformed problem to obtain an approximate solution to the original problem. In these approaches, if the objective function is not specified as a quadratic, then it is formed into a quadratic by a second order Taylor's approximation. The nonlinear econometric model is linearised by a first order Taylor expansion. Linear quadratic optimal control techniques are then used on the approximate problem (Kendrick, 1981). Frequently a fixed time horizon is specified, and the dynamic programming technique, which is based Bellman's principle of optimality, used to determine an optimal control trajectory. The main issues considered are, normally, the stochastic nature of economic models, and forward looking expectations variables. Rarely are computational issues, such as the effects of the truncation errors on the quality of the approximate solution in relation to the originally specified problem, considered.

These approaches have generated considerable research. They continue to do so, with, for example, Amman, Kendrick and Neudecker (1996) considering the approach with an infinite time horizon.

6.2.1 The OPTCON Algorithm

OPTCON (Neck and Matulka, 1994; Neck, 1996) is an algorithm for generating optimal controls for nonlinear macroeconometric models with uncertain parameters and additive model uncertainties. The basic approach is to find a fixed point for the trajectory of controls in repeated linearisations of the model, to form, for each iteration, a stochastic linear quadratic optimal control problem.

The OPTCON algorithm is concerned with time-invariant econometric models of the form

$$\mathbf{x}_t = \mathbf{f}(\mathbf{x}_{t-1}, \mathbf{x}_t, \mathbf{u}_t, \theta, \mathbf{z}_t) + \epsilon_t \tag{6.1}$$

Here $\mathbf{x}$ is a vector of states; $\mathbf{u}$ is a vector of controls; $\mathbf{z}$ is a vector of non-controllable exogenous variables; θ is a vector of uncertain, but with known mean and variance, parameters; and, ϵ is a vector of white gaussian noise which is independent of θ. The model is in discrete time, so that t represents a discrete sample point. There is a fixed time horizon from t_s to t_f. Models of the type specified by Equation 6.1 are the usual form for nonlinear macroeconometric models.

The objective function is of the finite time horizon, quadratic, state tracking type. The deviations between the actual and desired controls and states have fixed weights, but, following the usual economic practice, are discounted with time. That is, the objective function is

$$L = \sum_{t=t_s}^{t_f} L_t(\mathbf{x}_t, \mathbf{u}_t) \tag{6.2}$$

with

$$L_t(\mathbf{x}_t, \mathbf{u}_t) = \frac{1}{2} \begin{bmatrix} \mathbf{x}_t - \bar{\mathbf{x}}_t \\ \mathbf{u}_t - \bar{\mathbf{u}}_t \end{bmatrix}^T W_t \begin{bmatrix} \mathbf{x}_t - \bar{\mathbf{x}}_t \\ \mathbf{u}_t - \bar{\mathbf{u}}_t \end{bmatrix} \tag{6.3}$$

Here $\bar{\mathbf{x}}$ and $\bar{\mathbf{u}}$ are the given "ideal" (Neck and Matulka, 1994, p.385) levels of the states and controls, respectively. The weighting matrix, W_t, is partioned so that the sub-matrices weight the states and control appropriately, and is assumed to be symmetric. Further

$$W_t = \alpha^{t-1} W \tag{6.4}$$

so that the weights are only time varying because of the discount rate, α, which is itself constant.

This objective function is the usual form used with economic applications of linear quadratic optimal control theory. It is the discrete analogue of the objective function used by Currie and Levine (1993*c*) and considered in Chapter 2 (Equation 2.2 on page 18).

The OPTCON algorithm works in the following manner. Firstly, an initial guess is made of the optimal trajectories for the states and controls with the model solved by the Gauss-Seidel algorithm (Burden and Faires, 1985), using estimates for the states and controls, the expected values of the stochastic variables, and the known values for the other exogenous variables. The model is then linearised around this reference trajectory. This is a standard approach (Bryson and Ho, 1975; Chow, 1981*b*), but requires the calculation of derivatives for each function in the model with respect to each variable. The system matrices of the linear system are not only time varying, but are also random as they are a nonlinear function of the stochastic parameters (θ). The influence of the stochastic model parameters on the linear system matrices is calculated by a linear approximation of this function. This requires the calculation of the derivatives of the parameters of the linear system with respect to the original model parameters. These are effectively second derivatives for the original model. All derivatives are evaluated for the reference trajectory. So, too, are expected values, which is only correct, as the authors point out, if the system matrices are linear functions of the model parameters.

The problem is now turned into a stochastic, time varying, linear quadratic, state tracking one. This problem is then solved using Bellman's principle of optimality, *viz.*

$$J_t(\mathbf{x}_{t-1}) = \min_{\mathbf{u}_t} \mathcal{E}_{t-1}[L_t(\mathbf{x}_t, \mathbf{u}_t) + J_{t+1}(\mathbf{x}_t)] \tag{6.5}$$

Here J is the performance index and $\mathcal{E}(.)$ is the mathematical expectation operator. Using Bellman's principle of optimality, the optimal feedback rule is obtained.

This part of the algorithm uses the usual economists method of solving linear quadratic optimal control problems. An important issue with this class of problems has been the learning about the system parameters using adaptive control techniques (Kendrick, 1981). The OPTCON algorithm rules this out by using pre-knowledge of the stochastic parameters (i.e. initial values of expected values and covariances). But it is extended in Gaschler (1996) so that the means and covariances are time varying.

From the optimal control rule a new control trajectory for the time horizon is generated. The model is again solved, using the Gauss-Seidel algorithm, giving a new reference trajectory. This whole process is iterated until a fixed point is found for the control and state trajectories. The fixed point trajectory is then said to be an approximation to the optimal solution for the nonlinear model.

The strength of the OPTCON algorithm is that it takes the problem of developing economic policy with nonlinear econometric models expressed in their usual form, and transforms this problem into the more tractable problem of stochastic linear quadratic optimal control. It extends standard methods into nonlinear econometric models. But, there are a number of issues that can be raised about the algorithm. Firstly, given that derivatives are calculated, it is surprising that they are not used in the solution of the nonlinear model. A Newton solution is generally faster than the Gauss-Seidel solution algorithm. This may be because derivatives are calculated numerically and numerical differentiation has large truncation errors and can easily accentuate noise (Burden and Faires, 1985), but these derivatives are used in the development of the optimal control.

A more important issue is the question as to what is the found fixed point trajectory – is it the minimum, or even approximate minimum, to the originally specified problem? Just because successive iterates do not change by more than some prescribed tolerance does not imply that the iterate is the minimum. Approximating the nonlinear system by a first-order Taylor series introduces truncation errors. Continuously repeating this process at each iteration, may result in growth in these truncation errors. Further, the linear approximation is only appropriate in a near neighbourhood of the original trajectory, but the aim of the algorithm is to move away from this trajectory. The linearisation approach is usually used for small deviations around some reference trajectory (Bryson and Ho, 1975).

The algorithm is computationally complicated. It uses a fixed point iterative scheme within a fixed point iterative scheme. Many variables are stochastic. Besides the truncation error associated with the linearisation, it has the possibility of round-off errors through, for example, the calculation of complicated inverses. It is possible that numerical errors could significantly effect the convergence properties of the algorithm. The authors, though, point out that the algorithm only generates approximate solutions to the optimal control; but an important issue is how good are these approximations?

These points are applicable to most studies that use control techniques with economic models. The emphasis is usually on the economics and econometrics and usually less rigour is used in the control methodology. Less confidence can be placed in the results in these cases. On the other hand, it could be argued that the techniques for developing economic policy are not important as long as the policy gives the desired results.

In the application of their algorithm, Neck and Matulka (1994) run a set of experiments based on two similar low-order quarterly econometric models for Austria. These models are based on the IS-LM framework and are similar to those of Chapter 3 (Section 3.3 on page 67). The first model is a nonlinear model and is estimated using ordinary least squares, and the second is a linear model estimated with Three Stage Least Squares.

A set of historical simulation exercises are undertaken with the models. The essential aim of these experiments is to attempt to reduce the business cycle around smooth growth paths, and can be considered as removing fluctuations around some operating trajectory. For desired outputs initial values are either constant or grow at constant rates. The deviations from desired states and controls are given weights with, for example, GDP weighted at 10 and inflation weighted at zero. These are not discounted with time.

The deterministic case is considered, followed by the case with only additive model uncertainty, as well as the full stochastic case. The results show that the business cycle fluctuations can be reduced, though even for GDP in the deterministic case for the linear model, the desired trajectory is not attained.

These experiments are of the re-creation of history type experiments. The algorithm has also been used to develop optimal policies for Austria for the future (1995 to 2000) (Neck, 1996; Neck and Karbuz, 1996). The econometric model used in these experiments is similar to those above. Forecasts of exogenous variables are made using time series techniques. In these experiments they find that the optimal policies are not significantly effected by differing desired trajectories. They also find that global factors which effect Austrian export and import prices are important.

These applications of the OPTCON algorithm have been in the practical development of current macroeconomic policy for Austria. They show how small nonlinear macroeconometric models can be used in economic policy development using standard optimal control techniques. They illustrate the approximation approach to optimal control with nonlinear practical econometric models. They also illustrate the current active research and allow for a number of the issues to be highlighted.

6.2.2 Model Reduction and Linearisation

One other research group has combined model reduction with linearisation. This approach with large nonlinear econometric models has been to generate simpler linear models that, for the subset of endogenous and exogenous variables of interest, faithfully reproduces the nonlinear model. The techniques of model reduction and linearisation by estimating derivatives by perturbation are used (Kailath, 1980; Christodoulakis, 1989*a*). The approach has been used in conjunction with the London Business School Model by Christodoulakis et al. (1991) to consider the benefits of simple feedback rules for generating economic policy (Currie and Levine, 1993*c*).

The approach has also been used by Christodoulakis, Currie, Garratt, Kemball-Cook, Levine, Barrell, Ireland and Westaway (1993) to produce a lower order linear version of the massive Global Economic Model (GEM). They refer to the linear version as MINIGEM. The methodology is as follows. Firstly the number of countries is reduced to three blocs with the first being the U.S.A.; the second being Germany, U.K., France and Italy; and the third being Japan. Secondly, the outputs considered are real GNP, an inflation measure and the ratio of current account balances to GNP, giving 9 output variables in total. Thirdly, exchange rates and forward looking expectations variables were exogenised for GEM. Fourthly, GEM was then linearised through perturbations with the shocks given to government expenditure, real interest rates, exogenous commodity prices, exogenous oil prices and to real exchange rates of the other blocs. Fifthly, equations for exchange rates were added to MINIGEM based on Uncovered Interest Parity. Via these equations forward looking expectations variables were included. The resultant model was validated against the parent model.

For the control type experiments MINIGEM was put into a "stochastic realisation in deviation form" (Christodoulakis, Currie, Garratt, Kemball-Cook, Levine, Barrell, Ireland and Westaway, 1993, p.543). This gives a standard control form discrete linear stochastic state space model (i.e. the system {A,B,C,D}), but with the forward looking expectations states not being recursive. This is the type of model considered in Chapter 2. The control experiments were part of a wider exercise (Bryant and Mann, 1993) where a specified (quadratic) objective function and specified feedback rules were evaluated. Optimal control rules of the approach of Currie and Levine (1993*a*) were also developed and compared to the other rules. The Currie and Levine (1993*a*) approach to linear quadratic optimal control was considered in Section 2.2 (on page 17).

The MINIGEM approach develops a lower order linear model and uses that model for linear quadratic optimal control exercises. The loop between the lower order child model and the parent model is not closed. The parent model is used only as a validation of the child model. From this book it could be argued that a useful extension would have been to close this loop through output-injection and use MINGEM as an observer for GEM. As has been shown in earlier Chapters, the child model could then be used to learn the forward looking expectations variables, and, also, for the development of the control policy.

One of the problems of this approach is that it has generated a separate model. When different policy issues need to be considered or the parent model evolves, then the child model must be built again. Another problem is that the states of the child model (except those added by the exchange rate equations) may not have an economic meaning. It may then be difficult to implement or explain state feedback rules to policy makers. On the other hand, it allows for optimal control exercises to be undertaken using the information from what is a massive econometric model.

6.3 Nonlinear Optimisation

An alternative to linearising the nonlinear econometric model and using linear quadratic optimal control techniques, is the approach of using nonlinear optimisation with the model.

The problem is that of a dynamic constrained nonlinear optimisation. A common approach is to stack the problem up over the time horizon so that it becomes one large optimisation problem, and then use an optimisation algorithm to find the optimal control. This is an open loop approach rather than the closed loop approach of the linear quadratic approximation. An example is the work of Hall and Stephenson (1989) (Hall and Stephenson, 1990*a*; Hall and Stephenson, 1990*b*; Cividini and Siviero, 1992) where an algorithm is developed to allow for the stochastic nature of econometric models.

Two recent applications are considered below with the first emphasising forward looking expectations, and the second being an Australian study using essentially the same econometric model as used in this book.

6.3.1 Rational Expectations Models

In a study on the implementation of rational expectations on large scale econometric models, Fisher (1992) considers optimal control generation as a dynamic game between the policy makers and economic actors forming their rational expectations. This is a game in expectations forming rather than a game in control (Başar and Bernhard, 1991). Fisher (1992) proposes three algorithms all of which use the stacking and nonlinear optimisation approach. The first algorithm is a non-cooperative Nash game and is time consistent and expectations consistent. A time consistent solution is one where the policy maker has no incentive to change the optimal solution with the passage of time. The algorithm is a fixed point iterative solution to finding the expectations consistent optimal policy. That is, an initial estimate is made for the trajectory of expectations, then optimal control is found by some nonlinear optimisation routine using these expectations. This generates a new set of model consistent expectations, and the process is continued until (hopefully) convergence to the fixed point trajectory of expectations.

The second algorithm is a cooperative Nash game where policy makers stick to their announced policy and economic actors behave as if their expectations are inconsistent. This is the type of game where the objective functions are jointly minimised.

The third algorithm is a time inconsistent but expectations consistent solution. It is a Stackleberg (leader-follower) type game, and the policy makers' objective function is minimised with the expectations being model consistent for each period. In the implementation of this algorithm an attempt was made to combine the finding of the model consistent expectations with the finding of the optimal control. But convergence for expectations is usually dependent upon the controls being stable near the terminal time period, so the control problem was limited to a subset of the time horizon. This required modification of the usual (Newton based) solution to the stacked optimisation problem and required the heroic assumption that the effects of controls are not dependent upon the states.

As part of this study, Fisher (1992) uses the time inconsistent expectations consistent algorithm to investigate the inflation-unemployment tradeoff. He uses three UK econometric models each of which contain forward looking expectations variables. The Liverpool Model (LPL) is an annual model which has a new classical structure with a set of long-run equilibrium equations determined by the supply side. The London Business School Model (LBS), which has, since 1985, had three expectations variables in its financial sector, is a quarterly model based around the income-expenditure framework, and is considered a monetarist model. The National Institute of Economic and Social Research Model (NIESR) is a quarterly model which introduced eleven expectations variables in 1985, and is generally Keynesian in nature.

Fisher (1992) is concerned with two outputs of inflation and unemployment; and uses the two controls of government spending and income tax rates (both fiscal policy). The objective of the policy is to move the economy to a position with 0% p.a. for inflation, and an unemployment of 500,000. The base trajectories (i.e. normal model solution) were used for the desired values of controls. Weights are used for the deviations of the outputs from this desired position; deviations of the controls from their desired position; and for changes to the controls. The relative preference between inflation and unemployment is altered by changing the output weights, but keeping the relative preference of controls and outputs the same. Five different sets of relative weights are used; the neutral (that is, the initial weights of 1.5 for inflation and 2.7×10^{-5} for unemployment, 2.7×10^{-5} for government expenditure and 2500 for income tax); these scaled up and down by a factor of 3; and the neutral also scaled up and down by a factor of 19.

Using the LBS Model, the objective function was minimised for the first 20 quarters of the 32 quarter time horizon, and the results for inflation and unemployment were averaged over the last 4 (of 20) quarters. Using the five sets of relative weights, gave five points on the Phillips curve.

With the NIESR Model there was some convergence problems with some sets of weights, but the results obtained produced a much flatter Phillips curve than the LBS Model. On the other hand, the LPL Model produced contradictory results to the other models, with a Phillips curve of positive slope.

Fisher (1992) also experimented with his first algorithm which is a time consistent, expectations consistent algorithm. Using this algorithm with the inflation-unemployment trade-off problem, the NIESR and LBS Models converged in one step, so there was no incentive to re-optimise. That is, the problem was not time inconsistent. The LPL Model, though, failed to solve.

This study has looked at a practical economic policy issue using control techniques. The study has highlighted the importance of using a variety of models, for very different results were obtained from different nonlinear econometric models. The study has also highlighted some of the problems with the nonlinear optimisation approach, especially in regard to the forward looking expectations of economic actors.

6.3.2 Australian Studies

Recently two sets of studies have been undertaken using nonlinear optimisation with Australian models. The first of these (Murphy and Brennan, 1993; Murphy and Brennan, 1994; Murphy, 1996*b*) has been undertaken on versions of the Murphy model, and, as this book uses a version of this model (see Chapter 7), these studies are considered in detail. The second study (Louis, 1996) was undertaken at the Australian Treasury using their econometric model. Both the model and the optimal control application are similar to the Murphy study.

The Murphy Study

The Murphy model contains forward looking expectations variables that are solved using rational expectations techniques, but these are not emphasised in the Murphy approach as they were in the Fisher (1992) approach. The Murphy approach produces a time inconsistent, expectations inconsistent result. It also, unlike the Hall and Stephenson (1990*a*) approach, essentially ignores the stochastic nature of the model.

The Murphy (1996*b*) approach is as follows. Firstly, outputs and controls are selected and the model is simulated to produce a numerical result (look-up table) which is written as

$$\mathbf{Y} = \mathbf{h}(\mathbf{U}) \tag{6.6}$$

where $\mathbf{Y}$ is a vector of outputs for the control exercise stacked up for the entire time horizon, $\mathbf{U}$ is a stacked up vector of controls. The aim is to find the optimal controls to minimise an objective function. As with Fisher (1992), the controls are not free over the entire time horizon, but are restricted, for each of the control variables, to a single value for the second half of the time horizon. "This is to avoid the well-known problem of myopic behaviour towards the end of finite horizon optimal-control problem" (Murphy and Brennan, 1993, p.12).

The objective function can be written as

$$J = \mathbf{f}(\mathbf{Y}) \tag{6.7}$$

In Murphy and Brennan (1993) this is a quadratic, whereas in the more recent studies (Murphy and Brennan, 1994; Murphy, 1996*b*) it is of the quadratic type but with some weights being altered. Specifically, tax changes are restricted to the annual budget time by applying large weights to tax changes in all except the appropriate quarter; and, a zero weight is applied if the unemployment rate is below the target value.

Newton's method is used to minimise Equation 6.7. Derivatives for Equation 6.6 are calculated numerically, and it is argued that linearity can be assumed (for Equation 6.6) so that these need only be calculated once. (This begs the question of why not use linear techniques.) In the studies where the objective function is not a fixed quadratic, the first and second derivatives of the objective function are evaluated at each iteration. Analytic derivatives are used.

For a stochastic nonlinear model this approach could introduce large numerical errors. This may be compounded by the method of handling forward looking expectations. This could be important for, strong policy recommendations are made from these results, and no analysis is undertaken to issues such as the robustness of the results to numerical issues. This compares to the rigorours econometric analysis in the development of the Murphy model (Powell and Murphy, 1995).

In the first study (Murphy and Brennan, 1993) the first choice of controls is to use short term interest rates (90 day bank bills) and tax on labour income (the Murphy model variable POL1T) as their monetary and fiscal instruments. But both these variables are endogenous to the model, and so the relative change in the money supply ($\Delta M_t/M_{t-1}$) is used and it is argued that the money supply (M) has a one to one relationship with short term interest rates (RCS) in the model. For the fiscal policy instrument it is argued that "... the optimal policy response is unlikely to be very sensitive to the choice of instrument" (Murphy and Brennan, 1993, p.7), and so net public sector borrowing requirements (RPUBDE) is used.

Following most economic studies, the desired output is a point rather than a trajectory. The first output variable is the unemployment rate (UR), and the desired value is chosen to be 7.3%. The second output is an inflation rate measured as the percentage increase in average weekly earnings. A desired value of 4% p.a. for the Wage Inflation Rate (WINF). The last target is that of public debt as a percentage of Gross Domestic Product (B) and a rate of 33%, which is the current rate, as "... we are not aware of any study of the optimal level of public debt in Australia from which to obtain a better estimate" (Murphy and Brennan, 1993, p.9). These values just happen to be the steady state values to which the model is designed to converge.

The objective function also penalises changes in the controls to restrict rapid changes in policies. The weights are also subjectively chosen. The time horizon is 48 quarters, and future values are discounted. The objective function is given by

$$J = \sum_{t=1}^{48} \frac{1}{(1.0125)^t}[0.04(\mathrm{B}_t - 33)^2 + 4(\mathrm{UR}_t - 7.3)^2 + (\mathrm{WINF}_t - 4)^2 + 4(\Delta\mathrm{POL1T}_t)^2 + (\Delta\mathrm{RCS}_t)^2] \tag{6.8}$$

Using this objective function Murphy and Brennan (1993) undertake two sets of experiments. Firstly, they reconstruct policy for the 1980s in an attempt to estimate if policy could have been improved; and, secondly, they make some predictions about future policy. The main conclusions they make from this experiment are: "The optimal macro policy brings less variability in unemployment, and lower overall inflation compared with the actual policies of 1980-1992" (Murphy and Brennan, 1993, p.37); and that fiscal policy acts more quickly on economic activity than monetary policy.

The second experiment is to undertake future predictions based on the optimal policy. Using a baseline solution of a best guess at future changes to exogenous variables, they compare the optimal policy using the above objective function; one where unemployment and inflation are equally weighted (WINF's weight is increased to 4); and another with the target for inflation at 6.1% p.a.. With each of these the output slowly converges towards a steady state at its desired value after 48 quarters (i.e. the year 2004). Each result is closer to its desired position than the baseline solution.

In the most recent study (Murphy, 1996*b*), the inflation and unemployment trade-off is considered. This study uses the modified quadratic objective function and an updated version of the Murphy model (Murphy, 1996*a*) where short term interest rates have been exogenised and thus are used as a control. The desired values are again selected to be the steady state solution of the model, and experiments are undertaken on a predicted baseline model solution from 1995 to 2002.

In the first experiment the desired inflation target is reduced by 1%. In this study the GDP price deflator is taken to be the measure of inflation, and it has a steady state value of 3.5% p.a.. The study shows that inflation can be reduced at the cost of tighter monetary policy (higher short term interest rates) and a temporarily higher unemployment rate. The changes to unemployment are temporary as the model always returns unemployment to its steady state value. A slightly expanded fiscal policy is also recommended by this controller.

The second experiment is an interesting one. Here a shock is applied to the model by altering the wage equation so that the steady state unemployment rate, which is the non-accelerating inflationary rate of unemployment (NARIU), is increased. This results in wage inflation and an increase in real wages as the new NARIU is reached. The controller recommends a tightening in monetary policy to dampen the resultant inflation, and a tightening in fiscal policy (tax rates) as the tax base is eroded due to the unemployment.

The Treasury Study

Another Australian nonlinear optimisation study has been undertaken by the Australian Treasury. The study (Louis, 1996) uses the Australian Treasury model (TRYM) and is essentially the same as the Murphy studies. The TRYM model is very similar to the Murphy model, and the control study was strongly influenced by the Murphy and Brennan (1993) study. Again a quadratic loss function is specified, but, being the Treasury, a separate control is introduced for fiscal policy revenue and fiscal policy expenditure. Again, the time inconsistency problem is assumed away, so that the effects of the expectations of forward looking economic actors are assumed, for the control exercise, to be unimportant. The objective function differs, though, in that the desired output point is not the steady state for the model. The stacking over time approach is not used. A Newton algorithm is used to minimise the objective function subject to the model. Again, derivatives are numerically estimated and the model is assumed to be linear so the derivatives only have to be evaluated once.

The TRYM model, and so, too, the Murphy model, has a steady state or Baseline solution.

An experiment is undertaken to compare the optimal control regime with the standard Baseline result when the model is shocked exogenously. The shock used is an improvement in the terms of trade by a 5% increase in the price of commodity exports for a 4 quarter period. The results show that the controller dampens the oscillatory response of the model to the shock compared to the Baseline solution.

Relevance

The methodology and rigour of these Australian studies differ greatly from the approach used in this book. Perhaps the most interesting point is that they assume the model is linear, but use nonlinear techniques. The studies are important though in that they suggest a role for control techniques in the development and evaluation of economic policy. The importance of these studies is that they were designed to be practical and to influence Australian economic policy. Murphy, who is considered Australia's premier macroeconomic model builder, says that "optimal control is the appropriate tool for designing active macroeconomic policy" (Murphy, 1996*b*, p.159). Hopefully this will help in the implementation of the ideas in this book. One of the restrictions noted by Louis (1996) is with software. Again, the visual simulation approach and ability to easily modify weights and desired trajectories used in this work may help overcome such limitations.

6.4 Separate Model Based Control

The studies considered above are recent, yet the final study (Zarrop et al., 1979) considered in this Chapter was undertaken 20 years ago. It is examined here because it incorporated many of the features used in this book. It uses a lower order linear model to act as a controller for a large nonlinear econometric model. Unlike the Christodoulakis, Currie, Garratt, Kemball-Cook, Levine, Barrell, Ireland and Westaway (1993) study considered above, the study closes the loop between the low order model and the high order model.

The authors argue that the larger model is an analogue of the real economy and that it can be used to test their controller.

The London Business School model was used to represent the real economy. This is an earlier version of the same model as was used in the Fisher (1992) study, and, at that time, was a large, nonlinear, deterministic model measured in levels. Zarrop et al. (1979) use a lower order, stochastic, state space model to form the controller. In this model variables are expressed in percentage changes. This lower order model has the form

$$\mathbf{x}_{t+1} = F\mathbf{x}_t + G_u\mathbf{u}_t + G_e\mathbf{e}_t + G_w\mathbf{w}_t \tag{6.9}$$

$$\mathbf{y}_t = H\mathbf{x}_t + D_u\mathbf{u}_t + D_e\mathbf{e}_t + D_w\mathbf{w}_t \tag{6.10}$$

where $\mathbf{x}$ is a vector of states; $\mathbf{u}$ is a vector of controls; $\mathbf{e}$ is a vector of exogenous variables; $\mathbf{w}$ is a vector of gaussian stochastic disturbances; and, $\mathbf{y}$ is a vector of outputs.

The approach of Zarrop et al. (1979) is as follows. They eliminate the stochastic disturbances by solving the output equation (Equation 6.10) for $\mathbf{w}$, and substitute this into the state equation (Equation 6.9), giving

$$\mathbf{x}_{t+1} = A\mathbf{x}_t + B\mathbf{u}_t + C\mathbf{e}_t + K\mathbf{y}_t \tag{6.11}$$

where

$$\begin{aligned} K &= G_w D_w^{-1} & A &= F - KH \\ B &= G_u - KD_u & C &= G_e - KD_e \end{aligned} \tag{6.12}$$

Optimal control is then developed using Equation 6.11. At each time period, the control developed from this model is applied to the higher order London Business School model. To allow for the different number of controls and outputs between the two models, the optimal control generated on the lower order model was disaggregated before being applied to the London Business School model; and the output from it was aggregated before being incorporated into the control model. Further, as the control model is in differences and the London Business School model is in levels, accumulation is undertaken before the control is applied to the London Business School model; and, differencing is undertaken before its output is incorporated into the control model as outputs and exogenous variables ($\mathbf{y}$ and $\mathbf{e}$ in Equation 6.11) in the control model.

Using this approach Zarrop et al. (1979) undertake an historical tracking exercise whereby the outputs from the low order control model (which are % changes) are moved to a fixed point. The time horizon is 1970 to 1974.

The approach by Zarrop et al. (1979) and that used in Chapter 8 are similar. In both cases a linear low order model is used as the basis for an observer for a practical nonlinear econometric model. One difference is that here a learning gain term is incorporated in the observer so that the emphasis of the difference between the two model's outputs can be altered.

6.5 Conclusion

This Chapter has reviewed a number of recent studies that have used optimal control with nonlinear econometric models. It has presented the current thrust of research in this field, and in particular its use in current research into Australian economic policy. The first set of studies used the approach of linearisation and linear optimal control techniques. This is the area of early research, and the studies considered here show how this research continues. The second set of studies used nonlinear optimisation in generating open loop controls. This is the approach of much of the current Australian research. Finally, a 1970s observer based study was also considered, for this is the approach of this book.

All of the studies have used practical econometric models, and considered practical economic policy issues. This is particularly the case for the Australian studies. They were undertaken by two of the major players (Chris Murphy and the Australian Treasury) in the practical development of Australian macroeconomic policy by modelling. They suggest, and even argue for (Murphy, 1996*b*, p.159), the use of control techniques for policy development.

The most recent studies have not emphasised the importance of expectations as did the studies of the 1980s and early 1990s. The rational expectations debate that had a large influence on the use of control theory in economics (Kyland and Prescott, 1977) appears to have lost some of its influence.

An important issue that can be seen from these studies is that the emphasis is on economics in what is essentially a combined control and economics study. The control techniques are not as rigoroursly considered as the econometrics. Further, computational issues are not considered in any of the studies. An aim of this book is to consider both the control and computational issues. One approach to examining numerical issues is to examine a wide range of experiments; and tools to easily simulate scenarios facilitate this. The techniques presented in this book are useful here.

Two of the studies used separate models for developing control policies. The study by Christodoulakis, Currie, Garratt, Kemball-Cook, Levine, Barrell, Ireland and Westaway (1993) developed controls on a lower order linear model and these were validated *ex post* on the main model. The early study of Zarrop et al. (1979) used an observer approach. This approach is further considered here where, in Chapter 8, linear observers are used with the nonlinear econometric model developed in Chapter 7.

Chapter 7

A PRACTICAL MACROECONOMETRIC MODEL

7.1 Introduction

An aim of this work is to use current practice econometric techniques and models, and to undertake policy analysis with such models using control theory techniques based on different observer type models. The economic system as represented by the econometric model and the control system that generates the policies are separate, though communicating, objects. Each object can use best practices from its source discipline. In this manner, for example, concepts such as the model augmentation of Chapter 4 can be used with the control system and not impinge on the model building practices used in the economic system. This will, hopefully, allow the ideas of this book to be more widely accepted for they will not be dismissed by statements such as: 'that does not make econometric sense'.

This Chapter considers the development of a practical econometric model to act as the economic system object. A major difficulty for this project was access to such a model in a form that was useful for the project. One problem is that major Australian econometric models are commercial undertakings so that both access and many of the model's details are not generally available.

This book uses the Murphy model of the Australian economy. This dynamic macroeconometric model is arguably Australia's premier model, and is widely used by government, business and universities. It has had considerable influence on the development of Australian economic policy. It is, thus, well-suited for use in this book. But it is a commercial product and software for running the model is only available in executable code form, and this software is not useful for the type of analysis undertaken in this book. Some details including equation specifications and coefficients are available (Powell and Murphy, 1995), and this Chapter considers how those details were used to build a version of the Murphy model for use in this project. Thus this Chapter is concerned with the solution of a specified and estimated model, and, while the techniques used here follow economic model solving practices, the software and approach differ substantially from that usually used with economic models.

7.2 The Murphy Model

The Murphy model is a widely used dynamic macroeconometric model of the Australian economy designed for policy analysis and forecasting. Following most current macroeconometric models, it is a discrete model based upon quarterly data. Details of the model can be found in Murphy (1988*a*), Murphy (1988*b*), Murphy (1990*a*), Murphy (1990*b*), and Powell and Murphy (1995). Here a number of salient features are considered with the aim of presenting an overall impression of the model.

The model is strongly based upon economic theory, with the major design criteria being to build a model which is economically consistent and consistent with economic theory. The next design criteria relates to the statistics of the model, and it was designed to be consistent econometrically and to relate to Australian data. Thus the economics comes before the econometrics.

The model has many similar features to those considered in Chapter 3, though it includes much more detail. In the model wages and prices are sticky; wage inflation is modelled by an inflation expectations augmented Phillips curve and includes a trend rate of growth of productivity; price inflation is based upon a weighted average of past inflation; demand pressures are modelled by the level and changes in the level of unemployment; rational expectations are used in the financial markets; uncovered interest parity theory is used to model the exchange rate; and, the long term interest rate is based on expectations theory of the term structure.

The structure of the model contains three key features for economic policy analysis.

> *First, prices and wages are slow to adjust to shocks so the model has short-run Keynesian properties. Second, financial markets are forward-looking and deregulated, so there is a floating exchange rate and market determined bond rate which are responsive to new information, adding another dimension to the model's short-run properties. Third, in the long run there is optimising behaviour in production, implying that the employment, pricing, investment, exporting and importing decisions of the business sector are treated in an integrated and consistent manner.* (Murphy and Brennan, 1993, p.13)

These features produce a nonlinear model with considerable dynamics, but, having sticky components, can have a slow response. The forward looking behaviour of economic actors in the financial markets requires that the solution of the model needs knowledge of future behaviour and not simply past behaviour.

The Murphy model is being continually updated and modified depending on the demand for its services. It has recently (Murphy, 1996*a*) been extended to include much greater regional and industry details. The version considered here is that found in Murphy (1990*b*) and Powell and Murphy (1995).

The model follows the current approach to smaller macroeconomic models, and contains nearly 100 equations. A sample equation from the model is given in Box 7.1 and presents the equation for Private Consumption. It includes the estimated parameters, and the meaning of the variable names. The Box shows that the equation is complex and nonlinear. This consumption function contains 28 variables of which three are exogenous, and a comparison with the consumption function of Equation 3.13 (on page 69), which has two endogenous variables and a single exogenous variable, shows the increased complexity of the Murphy model.

As with most macroeconometric models, the Murphy model divides the Australian economy into a number of sectors and models these sectors for flows of goods and services, and for

Equation S04: Private Consumption

$$\log(\mathrm{CON}_t) = \mathrm{C042} \times \log(\mathrm{COND}_t) + (1 - \mathrm{C042}) \times \log(\mathrm{CON}_{t-1} \times \mathrm{GR}_t)$$
$$+ 400 \times \mathrm{C043} \times [(\mathrm{LPCON}_t - \mathrm{LPCON}_{t-1}) - (\mathrm{LPCON}_{t-1} - \mathrm{LPCON}_{t-2})]$$
$$+ 100 \times \mathrm{C044} \times [(\mathrm{NT}_{t-1}/\mathrm{NTS}_{t-1}) - (\mathrm{NT}_t/\mathrm{NTS}_t)]$$
$$+ \mathrm{C045} \times (\mathrm{RLIP}_t - \mathrm{RL}_t) + \mathrm{Z\_CON}_t$$

where (Equation SO4a)

$$\log(\mathrm{COND}_t) = (\mathrm{C040} + \mathrm{C041} \times \mathrm{TF}_t)/\mathrm{C042}$$
$$+ 2/3 \times \log[(1 - \mathrm{POL1}_t) \times \mathrm{W}_t \times \mathrm{NT}_t + \mathrm{TRPUPR}_t + (\mathrm{POL14}_t - \mathrm{POL13}_t) \times \mathrm{GDP\$}_t]$$
$$+ 1/3 \times \log[\mathrm{M}_t + \mathrm{B}_t - \mathrm{ZPA}_t - \mathrm{ZP}_t/\mathrm{ED}_t + \mathrm{PYD}_t \times (\mathrm{KH}_t + \mathrm{KINF}_t + \mathrm{KF}_t - \mathrm{FKF}_t)$$
$$+ \mathrm{PEXC}_t \times \mathrm{KIF}_t]$$

Coefficient	Estimate	Standard Error
C040	−0.2766	0.0458
C041	0.004553	0.000824
C042	0.3786	0.0660
C043	−0.000272	0.000255
C044	−0.00557	0.00218
C045	0.002212	0.000664

Variable	Description (Variables are endogenous unless specified)
CON	private consumption
COND	equilibrium private consumption
GR	natural real growth factor
LPCON	log of constant utility price index for private consumption
NT	total employment
NTS	labour supply
RLIP	average interest paid on outstanding $A denominated debt
RL	10-year government bond rate
Z_CON	stochastic error term (exogenous)
TF	fixed time trend (exogenous)
POL1	rate of tax on labour income
POL13	money growth rate
POL14	transfers form foreign sector to private sector as a proportion of GDP
W	wage rate
TRPUPR	net transfers from public sector to private sector
GDP$	nominal gross domestic product
M	private sector holdings of currency (exogenous)
B	private sector holdings of bonds
ZPA	private sector foreign debt denominated in $A
ZP	private sector foreign debt denominated in foreign currency
ED	exchange rate for converting ZP to $A
PYD	price of domestic goods
KH	stock of dwellings
KINF	stock of private non-farm inventories
KF	private enterprise stock of business fixed capital
FKF	stock of foreign owned business fixed capital
PEXC	price of commodity exports
KIF	stock of farm inventories

Box 7.1: Murphy Model Consumption Function

financial activities. For example, the Business Sector has 35 equations; uses three input variables of labour, business capital and imports; and, produces the three output variables of commodity exports, other exports and domestic goods.

The model is based around the assumption of a stable steady state solution (long run equilibrium) and attempts to explain the adjustment process to this. This stable steady state is dependent upon optimising behaviour, with, for example, the Business Sector aiming to maximise profits. The model's short run dynamics are effected by, for example, the stickiness of prices, and the costs of adjustment (e.g. to the capital stock). This adjustment process can be slow, with the model taking 3 to 10 years to incorporate the full effects of a wage shock (Murphy, 1990*b*, p. I-17). Thus the response of the model can be slow.

The stochastic equations of the model are estimated using quarterly data. The model uses standard econometric techniques for parameter estimation and uses a battery of tests to examine their econometric properties. Most equations are estimated by Ordinary Least Squares; 4 with nonlinear least squares; Instrumental Variables are used for one equation; and, the 2 long run supply equations are jointly estimated. Thus nearly all the equations are single equation estimates. Some parameters of the model are calibrated. The parameters for the consumption function can be seen in Box 7.1.

7.3 Model Solution

The Murphy model software follows a common practice in solving econometric models (Challen and Hagger, 1981) by first determining a 'control solution' for the entire time horizon, and then solving the model for the entire time horizon to create the 'shocked solution'. The 'control solution' contains a complete solution for all exogenous variables; and these, for historical simulations, are usually the variables historical values. For model predictions, the exogenous variables must first be predicted, and this is, by definition, outside the model. A usual simulation scenario with the model is to examine the implications of differing values for a subset of the exogenous variables. Frequently multiplier analysis is undertaken whereby the response of endogenous variables of a step or impulse in an exogenous variable are considered. Another common policy simulation scenario is to re-create history by considering what the model determines are the effects if, *ceritus paribus*, some policy setting such as a tax rate were different. In this book, this question is turned on its head to ask: 'what policy settings are necessary to achieve some desired trajectory for some of the endogenous variables?'. The re-creation of history is always subject to criticisms of the Lucas (1976) kind where the *ceritus paribus* assumption, and even the model itself, may be no longer valid under the new scenario.

The concept of a 'control solution' being solved to generate a 'shocked solution' over the entire time horizon makes difficult the closed loop concepts used here. This and the need for commonality in software, are the reasons for the creation of software to solve the Murphy model.

7.3.1 Methodology

The Murphy model can be written as

$$\mathbf{X}_t = \mathbf{F}(\mathbf{X}_{t-i}, \mathbf{X}_t^e, \mathbf{Z}_{t-j}, \epsilon_t) \qquad \begin{aligned} i,j &= 0,1,2,\cdots,N \\ t &= 0,1,2,\cdots,T \end{aligned} \tag{7.1}$$

where t represents the discrete time period of a quarter within the time horizon $[0,T]$; N is the maximum lag length; $\mathbf{X}$ is a vector of endogenous variables; $\mathbf{X}^e$ is a vector of forward looking expectations variables; $\mathbf{Z}$ is a vector of exogenous variables; and, ϵ is a vector of error terms. Initial and lagged values of the endogenous variables are given; exogenous variables are known for the entire time horizon; the error terms statistical properties are known from the estimation procedures; and, forward looking expectations are model consistent and are predicted by the model.

The model has 96 endogenous variables and 90 exogenous variables. The equations of the model use, besides these variables for the current quarter, lagged exogenous and lagged endogenous variables. These lags vary from a single quarter to a year ($N = 4$). For a given quarter these lagged endogenous and lagged exogenous variables are predetermined, and by a suitable redefinition of variables, can be considered as current states of the model. The endogenous variables and the expectations variables can also be considered as states of the model. The model then consists of current states and current exogenous variables for the quarter. This gives a total of 258 variables for each quarter. The model is then in the form

$$\mathbf{x}_t = \mathbf{f}(\mathbf{x}_t, \mathbf{z}_t) \tag{7.2}$$

where $\mathbf{x}$ is a vector of state variables, and $\mathbf{z}$ is a vector of current exogenous variables. The model contains the three forward looking expectations variables of the expected exchange rate, the expected bond rate and the expected inflation rate. These variables are taken to be model consistent, that is the model's predictions. The expected bond rate and inflation rate are the model's prediction for the next quarter (Murphy, 1990*b*, p.III-3). For the solution method here, the model is solved with these predictions being initially estimated as the current quarter's values, and then the model is re-solved with these variables being the model's actual predictions.

Notice that the model as presented in both Equations 7.1 and 7.2 has contemporaneous variables on both sides of the equation. Here, the model is solved by the Gauss-Seidel fixed point iterative algorithm (Burden and Faires, 1985). This is a common method used with nonlinear econometric models (Challen and Hagger, 1981). It suffers from a number of problems common to fixed point schemes including the fact that convergence is not guaranteed, and it can be sensitive to initial estimates. Following common economic practice, historical data is used as the initial estimate for the solution.

In this manner, the model is in a form that, given a vector of exogenous variables for a given time period, and given an initial estimate of the endogenous variables for that time period, then the model solution for the endogenous variables can be determined. Effectively the model is passed a vector of 258 variables being the 'control solution' for the current quarter and returns the same sized vector being the 'shocked solution' for the current quarter. A subset of the endogenous variables are considered the model's output (or policy target)

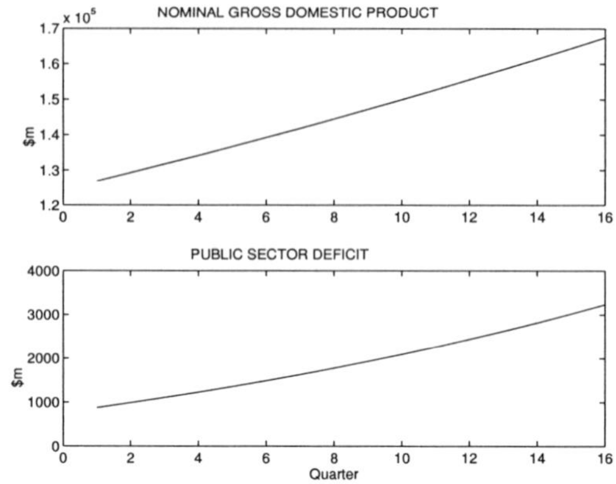

Figure 7.1: Baseline GDP and Public Sector Deficit

variables and a subset of the exogenous variables are considered the model's control (or policy instrument) variables.

7.3.2 Baseline Solution

The Baseline Solution for this dynamic nonlinear macroeconometric model is the steady state solution where all exogenous variables grow at smooth constant rates. Stochastic equation error terms are also held constant. The resultant trajectories of the endogenous variables will also be smooth. The Baseline Solution is useful for policy analysis as the effects on the model outputs of shocks to the control variables are clearly seen. It is also not limited to a specific time horizon as all exogenous variables can be easily extrapolated.

The technique for building the Baseline Solution was first to obtain smooth growth rates for all exogenous variables through the use of filters created by using ordinary least squares fits. Some of the variables were fitted to a linear model and others to an exponential model. The dynamic nonlinear macroeconometric model was then solved over a Baseline time horizon and the final half of the resulting series for each variable used to extrapolate the initial half of the series. The extrapolation was made using an ordinary least squares linear or

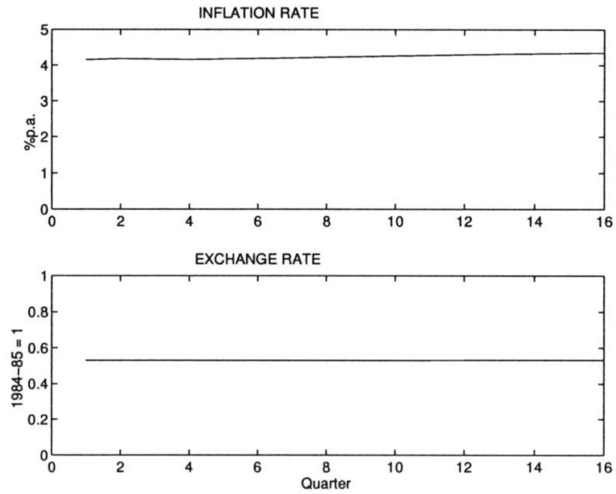

Figure 7.2: Baseline Inflation Rate and Exchange Rate

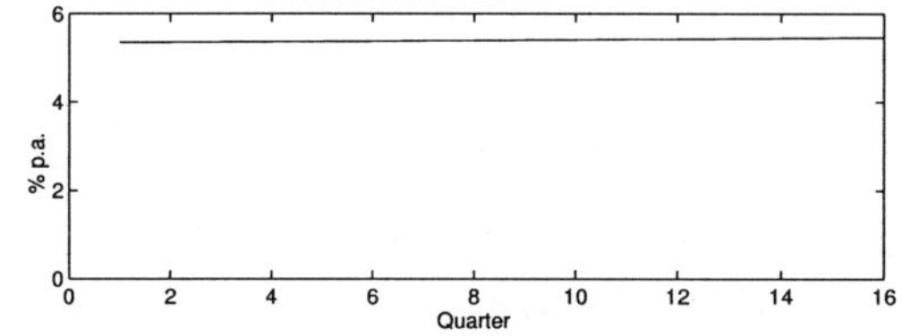

Figure 7.3: Baseline Interest Rate and Unemployment Rate

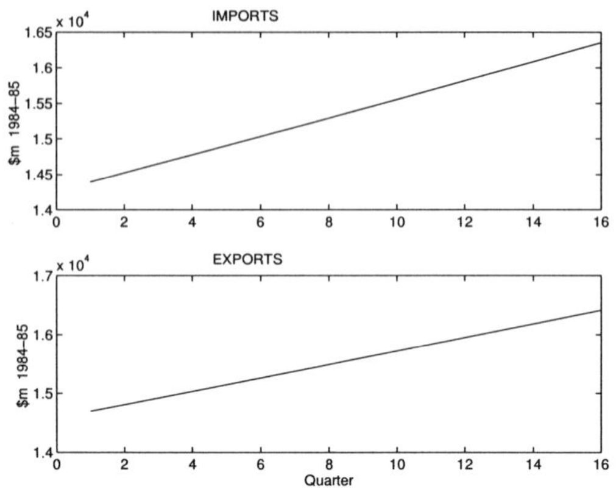

Figure 7.4: Baseline Imports and Exports

exponential fit. This process was iterated until convergence between the shocked solution and the control solution. Thus the technique for determining the Baseline Solution involves a fixed point iterative procedure to solve the model within the fixed point iterative solution to determine the growth rates for the exogenous variables. It is a computationally expensive procedure.

The procedure generated a set of exogenous variable data such that if the model is run over the Baseline time horizon with no change to any exogenous variable, then the resultant shocked solution endogenous variables will be (approximately) the same as in the control solution. That is, solving the model over the entire time horizon, with, for each time period, the Baseline estimates of the model's solution for the endogenous variables as the initial estimates, results in these estimates being correct. If an exogenous variable changes, then this will not be the case. The model is in a steady state, unless there is a change to an exogenous variable. Using this Baseline Solution, the model is particularly useful to examine the effects of a change in a control variable on an endogenous variable. That is, the multiplier relationship between controls and outputs.

The Baseline Solution for a number of key economic indicators are presented in Figure 7.1,

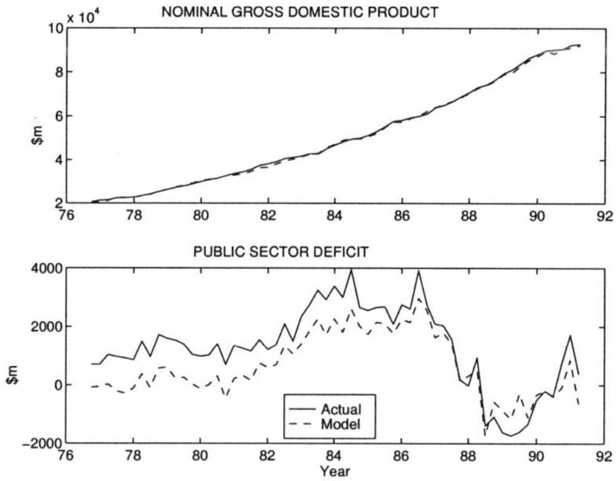

Figure 7.5: Historical GDP and Public Sector Deficit

through Figure 7.4. Figure 7.1 presents the trajectory of the endogenous variables gross domestic product and the public sector deficit over a Baseline time horizon of 4 years. Figure 7.2 gives the Baseline trajectories from the model for the inflation rate and the index for the exchange rate. Figure 7.3 presents the model outputs for long-term interest rates and the unemployment rate. And finally, Figure 7.4 presents the output trajectories for imports and exports.

7.3.3 Historical Solution

The Historical Solution uses actual historical data for exogenous variables and lagged endogenous variables, so that all model inputs and the initial conditions for the states are historical values. Here the model predicts what occurred historically. The time horizon is from 1976 quarter 3 to 1991 quarter 1. This time horizon contains the sample period for which the model was estimated[1], so that the Historical Solution is mainly a within sample period solution.

[1] Equations were estimated from 1976.1 to 1989.3 except for the money demand equation which uses the period 1980.1 to 1989.3 (Murphy, 1990*b*, p.I-21).

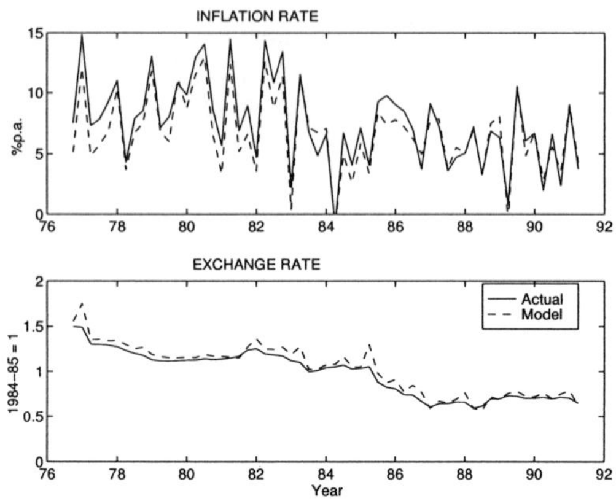

Figure 7.6: Historical Inflation Rate and Exchange Rate

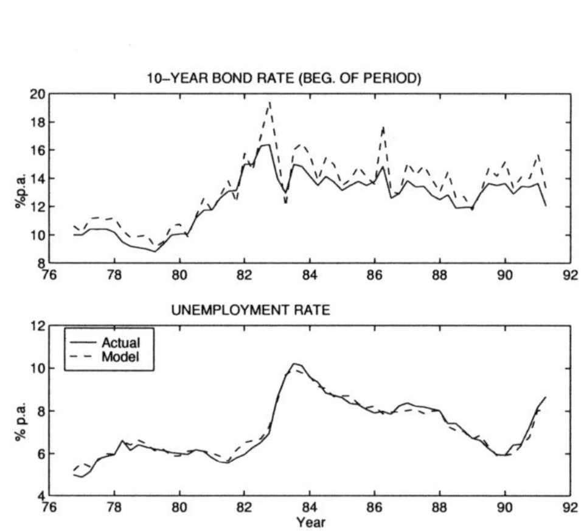

Figure 7.7: Historical Interest Rate and Exchange Rate

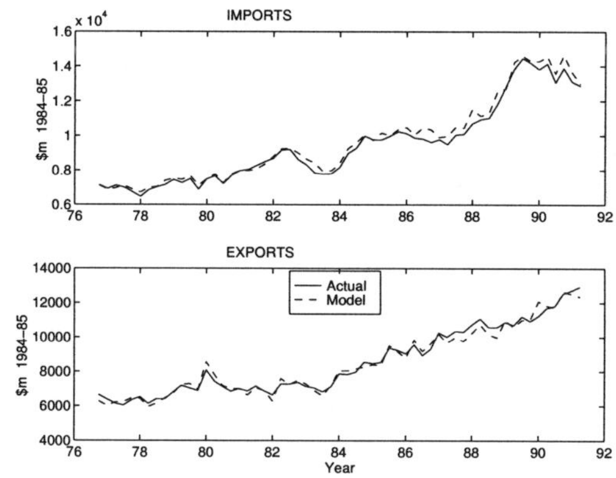

Figure 7.8: Historical Imports and Exports

The ability of the model to re-create the historical trajectories for the same output variables as in the Baseline Solution above, are presented in Figures 7.5 through 7.8. In these Figures the actual historical data for the variables is also plotted. Figure 7.5 shows that the model predicts GDP very well, and the public sector deficit reasonably well. In the latter case, most of the turning points are predicted well. Figure 7.6 shows that the model produces good estimates for both the inflation rate and the exchange rate, with both the levels and turning points being well predicted. Similarly, Figure 7.7 and Figure 7.8 show that the model is good at estimating long term interest rates, the unemployment rate, imports and exports.

It can be seen from these Figures that the model is a good model for replicating the historical data over this time horizon.

7.3.4 Numerical Issues

The Murphy model was designed with the strong aim of being consistent in economic theory, followed by the aim of using good econometric techniques. Numerical issues were not considered to be important in Murphy's original development of the model. Rather, errors in the model were felt to be introduced through miss-specification and estimation procedures. But once the model is defined there is also the problem of numerical errors introduced in its solution and simulation. Here, the model is taken as given, and as presented in Powell and Murphy (1995). This is not presented in a form that would minimise numerical errors, but rather in a form that makes structural economic sense. The functional form of many equations are such that slightly perturbed parameters could result in substantially different results. For example, adding the standard error to coefficient C043 in the consumption function given in Box 7.1 may significantly effect the estimate for private consumption. Equations of the model use terms that could introduce numerical problems with, for example, a common term being the log of the ratio of one quarter's value to the previous value. An equation often has many such terms. There are massive differences in magnitude between many data series, with some, despite being measured in millions of dollars, still being of the order of 10^4, and others being based on indexes around unity. Such data series are illustrated in the Historical Solution Figures. There are also massive differences in some parameter values. The model, as reflecting the real economy, is sensitive to small changes in the magnitude of many variables, and to small changes in the direction of many variables. It is possible that such changes could result simply from the numerical problems rather than from underlying changes. These round-off problems are likely to be compounded by the iterative procedures used in the model solution.

For the solution of the model here, care was taken so that these numerical problems would be minimised, and the software developed here is much less susceptible to numerical problems than that originally used with the Murphy model. In many cases this involved simple changes like the use of double precision for floating point numbers, rather than single precision as was used with the original.

7.4 Visual Simulation of the Model

The model was also developed into a form for visual simulation. Following the visual simulation techniques of Chapter 5, the model can be considered as an object within a

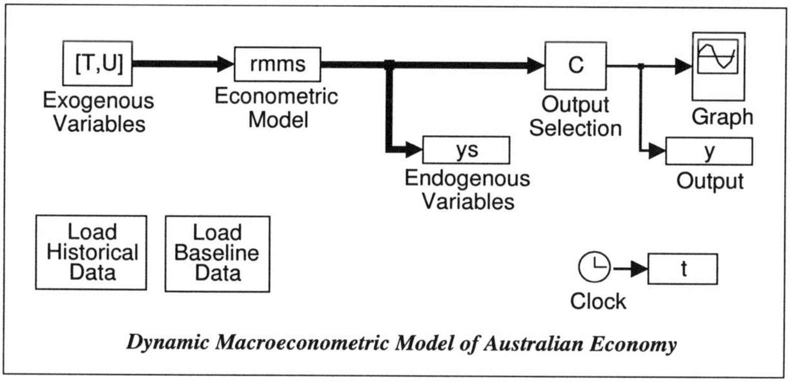

Figure 7.9: Visual Simulation Environment

visual simulation environment. The model was built as a high order object rather than being built up from low order objects, so that this form uses exactly the same source code as was used in the previous section. That is, exactly the same code for the model can be run in a visual or conventional simulation environment. This allows for the numerical properties of the model solution to carry over to the visual object for the model.

A visual simulation environment to run the model is presented in Figure 7.9. In the Figure, the *rmms* object solves the model for a single time period. Again its input is the control solution vector of predetermined variables (exogenous and lagged endogenous) and an initial estimate of the endogenous variables and it returns the shocked solution vector where the endogenous variables have been solved. As before, these variables can be considered as the current states ($\mathbf{x}$), and current exogenous variables ($\mathbf{z}$).

The *Load Historical Data* object in Figure 7.9, if selected, loads the data for the Historical Solution; whereas the *Load Baseline Data* object loads data for the Baseline Solution. The *[T, U]* table object contains a table of the states and current endogenous variables for each quarter over the entire time horizon. If the time step for the solution routine is set to 1 quarter, then within time horizon simulation uses data directly from the table. It also allows for shorter time steps, if this is considered appropriate, by interpolating data between

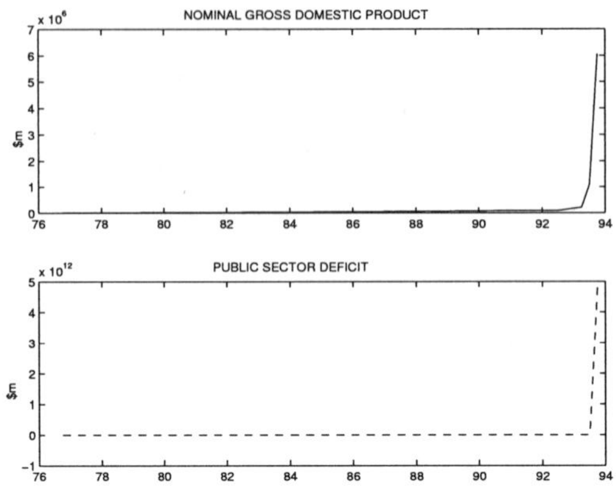

Figure 7.10: Extrapolated Historical GDP and Public Sector Deficit

quarterly table entries. Model prediction (i.e. simulation beyond the time horizon of table entries) uses extrapolation from the table.

The other objects used in the simulation environment of Figure 7.9 are of the same class as have been introduced in Chapter 5. The *ys* storage object is a data sink for storage of the states; the matrix object *C* selects some of what are considered the model outputs of interest (in the Figure a single variable is selected); these outputs are plotted as the simulation progresses by the *Graph* object; and, the outputs are stored by the *y* storage object for post simulation analysis. Again, the *Clock* object displays the simulation time and the *t* storage object stores it for post simulation analysis.

The simulation environment can be used to generate the same results as are presented in Figures 7.1 through 7.8. More complex simulation environments can be built up using these components, as will be seen in later Chapters.

As, in the Baseline Solution variables grow smoothly, the extrapolation properties of the *[T, U]* table object allow the model using Baseline Data to be simulated over as long a time horizon as is desired. The model can be run out indefinitely. This has the substantial benefit that changes can be made to exogenous variables during a simulation, and the effects examined visually. This is not the case with the Historical Solution. Here extrapolation, as would be expected, generates poor estimates of 'real' data and the model quickly explodes. This can be seen in Figure 7.10, which plots Gross Domestic Product and the Public Sector Deficit from the visual simulation with the time horizon extended for 10 quarters. That is, the exogenous data and initial estimates for the model solution are extrapolated two and a half years past the usual finish time. Figure 7.10 shows that the model quickly explodes. Figure 7.10 is directly comparable to Figure 7.5, and the estimates over the time horizon used in Figure 7.5 are identical. Besides highlighting the benefits of the Baseline Solution, this illustration shows the sensitivity of the model solution.

7.5 Conclusion

This Chapter has been concerned with the development of a high-order, nonlinear, practical macroeconometric model for use in this book. It is beneficial for the book to have such a model that uses best current practices within the econometric model building community. Such a model was not available in a usable form for the research here; but a listing of equations and coefficients of arguably Australia's major dynamic macroeconometric model, the Murphy model, was available. This Chapter has shown how a version of this model was developed for use in this book.

The Chapter considered the software solution of the model in a form useful for this book. This form differs from the usual manner used with macroeconometric models. In particular, from the normal manner for the Murphy model. Both the form and approach used to solve this model differs from that used in conventional econometric practice. Emphasis here was placed on computational issues in the model solution.

The Chapter has shown a Baseline or steady state solution for the model, as well as an Historical solution. In the Historical solution actual historical data is used for initial conditions and exogenous variables. Sample outputs have been used to show that the model performs well over this particular historical time horizon.

The Chapter has also shown how the model can be used in a visual simulation environment. This makes simulation of the model easier and helps in presenting the ideas developed in this book. The model is then in the form that it can be used as an *Economic System* object (as in Figure 5.7 on page 162) for simulation and policy analysis through control techniques. This is a novel approach to the use of econometric models.

Chapter 8

LINEAR OBSERVER BASED CONTROL WITH THE NONLINEAR MACROECONOMETRIC MODEL

8.1 Introduction

This Chapter is concerned with the use of linear observers to develop control policies for the practical, nonlinear, econometric model developed in Chapter 7. This is a large, with 258 states, dynamic macroeconometric model. The linear observers are based upon the models developed in Chapter 3. The first part of the Chapter focuses on control experiments, and the second part focuses on tools. In the second part, visual simulation environments are developed whereby policy design through control experiments can be undertaken. Again, using the techniques of Chapter 5, the objective is for practical experiments to be able to be undertaken by practical economists with the aim of disseminating the ideas of this book.

The first part of the Chapter uses, and extends, the ideas from earlier Chapters in presenting the results from experiments using low order, observer based controllers with the nonlinear econometric model. The experiments begin with simple, single-input, single-output observers in Section 8.2. These are expanded, in Section 8.3, to multiple-input, multiple-output observers that include the ramp tracking techniques of Chapter 4. These techniques are then used in conjunction with the visual simulation tools in Section 8.4.

The advantage of linear models is that the vast array of linear control techniques can be used. In particular in this Chapter, they allow for the use of linear quadratic optimal control theory to develop control policies to track desired output trajectories. Here, the control is developed on the linear model and applied to both the linear model and the nonlinear econometric model. The linear model is extended to an observer by incorporating a learning component through the injection of the nonlinear model's output. This is the approach used in earlier Chapters, and illustrated in Figure 8.13 (on page 231), but with now the *economic system* object being the nonlinear econometric model. Again, the aim is to have a clear division between the *economic system*, which is now a premier econometric model, and the *control system* which incorporates sophisticated control techniques.

The theoretical underpinnings of the use of observers is the Separation Principle. It requires both linearity and full and complete knowledge of the system to be controlled. With some missing knowledge (i.e. the models are effected by additive white noise of known variance) then the Certainty Equivalence part of the Separation Principal can be used. In the situation here, the Separation Principal is known not to apply for, firstly, the system is nonlinear and the controller is based upon a linear model; and, secondly, the linear model is of lower order than the econometric system. The linear model is not a reduced-order observer (Ogata, 1995), but a model developed independently from the system being controlled. This approach is used here so that the control system is distinct from the economic system for which it is developing policies. In this manner, and unlike most previous studies (see Chapter 6), the economic system is not modified or developed especially for the control experiments. Thus best practice econometric modelling can be used to produce the economic model, and the control system need not be constrained by econometric modelling practices.

Of course, in the real world the Separation Principle does not fully apply as linearity, and full and complete knowledge are not generally available. Here, for example, Taylor's theorem requiring the linear model to be in a near neighbourhood of the nonlinear model, is relied upon. It is for this reason that a wide range of experiments are undertaken. Thus a variety of models are used in this Chapter; and, more importantly, the simulation environments to undertake a variety of experiments are presented. This compares to most other studies where basically a single experiment is undertaken.

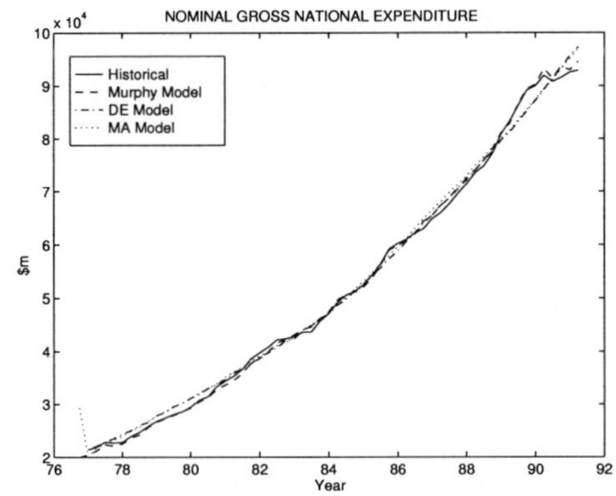

Figure 8.1: Simple Observers – Model Output

8.2 Simple Observers

The approach of the simple observer experiments is that of a deterministic linear quadratic tracking problem, and can be expressed algorithmatically as

$$\mathbf{x}_t = \mathbf{f}(\mathbf{x}_t, \check{\mathbf{z}}_t, \mathbf{u}_t) \tag{8.1}$$

$$\mathbf{y}_t = H\mathbf{x}_t \tag{8.2}$$

$$\hat{\mathbf{y}}_t = C\hat{\mathbf{x}}_t + D\mathbf{u}_t \tag{8.3}$$

$$\hat{\mathbf{x}}_{t+1} = A\hat{\mathbf{x}}_t + B\mathbf{u}_t + E - L(\hat{\mathbf{y}}_t - \mathbf{y}_t) \tag{8.4}$$

Equations 8.1 and 8.2 are the econometric model. The controls, $\mathbf{u}$, are explicitly identified as they are no longer exogenous, and an output equation (Equation 8.2) is specified with the matrix H selecting those states that are considered outputs. The econometric model is essentially the same form as Equation 7.2 (on page 199) but with the controls and outputs explicitly indicated. Equations 8.3 and 8.4 are the observer, and the output equation (Equation 8.3) is evaluated before the state equation as it uses the information content up to period $t-1$ (Åström and Wittenmark, 1990).

For these experiments, the closure of the system is through the optimal control rule where the control is generated by a discrete linear quadratic output tracking problem using the linear model. This type of problem was considered in Chapter 4 (Section 4.5 on page 142). Also for these experiments, the learning gain component, L, of the observer is chosen by the pole placement technique.

The first set of experiments are concerned with the level of economic activity and target the level of Gross National Expenditure. Pagan (1996), for example, argues that such variables are more important policy objectives than concentrating on objectives such as inflation and unemployment. The fiscal policy variable of government expenditure is used for the control.

Two single-input, single-output models are used as the basis for the observers. Both models were developed in Chapter 3: with the first being the Multiplier Accelerator model of Section 3.2 (on page 65); and the second being a contemporaneous control, second-order, difference equation of the type considered in Section 3.4.1 (on page 89). This latter model, which has the same inputs and outputs as the Multiplier Accelerator model and could be considered its reduced form, is given by

$$Y_t = \theta_1 + \theta_2 Y_{t-1} + \theta_3 Y_{t-2} + \theta_4 G_t \tag{8.5}$$

In Equation 8.5, Y is Gross National Expenditure, and G is government expenditure.

Figure 8.1 presents the output, GNE, from the models. It plots the historical data; the output from the dynamic nonlinear macroeconometric model developed in Chapter 7; the output from the Multiplier Accelerator (MA) model; and, the output from Equation 8.5, the difference equation (DE) model. The two linear models were estimated using the techniques considered in Chapter 3. The Figure shows that the models fit the data well. The estimated models are both controllable and observable.

The experiments undertake a counter-factual historical tracking exercise whereby the aim is to generate control policies for government expenditure that, if they had been introduced, would have reduced the business cycle in GNE. This, as with all counter-factual historical exercises, requires the *ceteris paribus* assumption; and all things may not have been equal if these policies had been introduced.

One advantage of generating policies to reduce the business cycle is that they can be considered as deviations about a trend. Thus the low order linear model can be considered to be operating in a neighbourhood of the nonlinear model, so that the Separation Principal is more likely to apply.

An issue where the models are of different order is the linkage between the models. In the nonlinear macroeconometric model, government expenditure is separated into two exogenous variables, with one, the larger, being expenditure on wages and salaries, and the other being general government expenditure (Murphy, 1990*b*). Here the latter is used as government expenditure in the control. This is particularly relevant to the Multiplier Accelerator model where the data series used in the nonlinear macroeconometric for consumption (CON, Private Consumption), investment (IBFT, Business Fixed Investment) and general government expenditure (GGCO, General Government Purchases of Consumption Goods) will not sum to produce GNE (GNEZ, Nominal Gross National Expenditure). Thus the national accounting identity ($Y = C + I + G$) in the model (Equation 3.3 on page 65) will not hold.

An advantage of the observer based approach is that it learns to respond like the system for which it is generating control policies, and thus may be able to adapt for such linkage differences.

For the experiments, the following scenario is constructed. It is assumed that the desired output is the ordinary least squares line through the Historical Baseline of the nonlinear macroeconometric model offset so the intercept is equal to the first observation. The Historical Baseline is the model solution when all exogenous variables take their historical values so the controller is not implemented. This desired output generates the objective of policy makers to reduce the fluctuations in the business cycle over the time horizon. The aim of the experiments is to consider whether such a desired output is achievable with the nonlinear macroeconometric model by adjusting general government consumption, and what are the implications if such a policy had been introduced, with all other things being equal.

The control parameters for the experiment are as follows. The control rule is developed

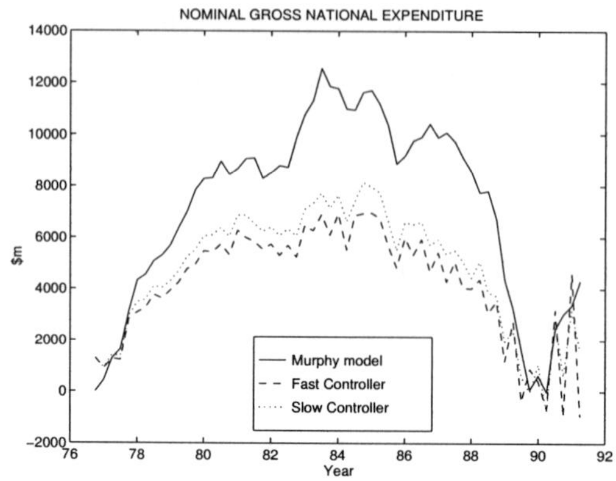

Figure 8.2: Multiplier Accelerator Model Output - Deviations from Desired

using a stationary linear quadratic regulator on each of the linear models. The output is considered more important than the input. This is accomplished by using weights of $Q = 3$ and $R = 1$ in the objective function. Further, desired input is chosen such that the linear model would achieve the desired output. For each observer, two learning gains are used with the first being faster (the eigenvalues of $(A - LC)$ are zero and 10^{-7}) and the second being slower (the eigenvalues of $(A - LC)$ being 0.5 and 0.55).

This scenario is basically using fiscal policy to consider economic output. It is not concerned with large changes in economic output, but more in smoothing the cycle. The nonlinear macroeconometric model does not allow fiscal policy to be adjusted at no cost, but has a policy reaction function (or control rule) whereby personal income tax rates are adjusted to force the ratio of the public sector deficit to GDP from not expanding beyond an exogenous desired level. It is probably for this reason that Murphy (1996*b*) uses this ratio (RPUBDE) as the fiscal policy control, and is a reason why it is so used in later experiments.

The output results for the Multiplier Accelerator model are shown in Figure 8.2. The Figure plots the levels as deviations from the desired output. The Historical Baseline, or uncontrolled, nonlinear macroeconometric model GNE where government expenditure takes

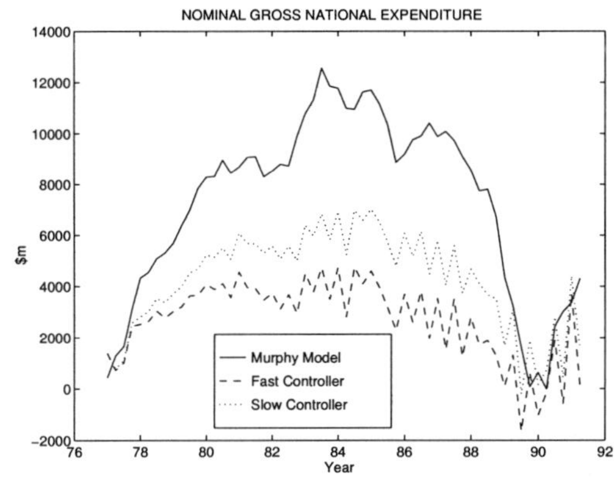

Figure 8.3: Difference Equation Model Output - Deviations from Desired

its historical values is plotted. The Figure also plots the results if government expenditure is developed by the controllers, with the fast controller using the faster learning gain. The Figure shows that using policies generated by the controller there is less deviation from the desired output. The faster learning gain produces slightly better results. The relative error (as measured by the 2-norm) between the actual and desired outputs is 12.7% with no controller, 7.5% with the faster learning controller, and 8.5% with the slower learning controller. These results show that generating government expenditure using the optimal control rule would have produced less variability in GNE than was historically the case with the nonlinear macroeconometric model, with all else equal.

The difference equation model produces the results presented in Figure 8.3. These results are better than those from the Multiplier Accelerator model with the relative errors being 4.9% and 7.4% for the faster and slower learning gains respectively. This model is a time series model and may be considered as a reduced form of the structural Multiplier Accelerator model. It is interesting that it produces better results, and it probably indicates the limitations of the Multiplier Accelerator model as an explanation of the business cycle.

Figures 8.4 and 8.5 plot the trajectory of general government consumption that was gener-

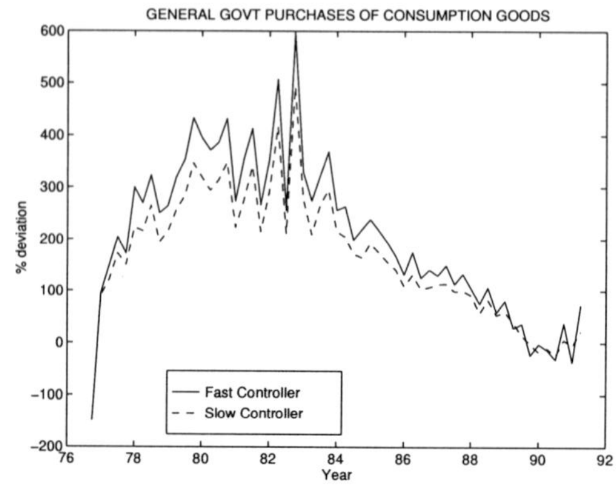

Figure 8.4: Multiplier Accelerator Model Control - % Deviations from Base

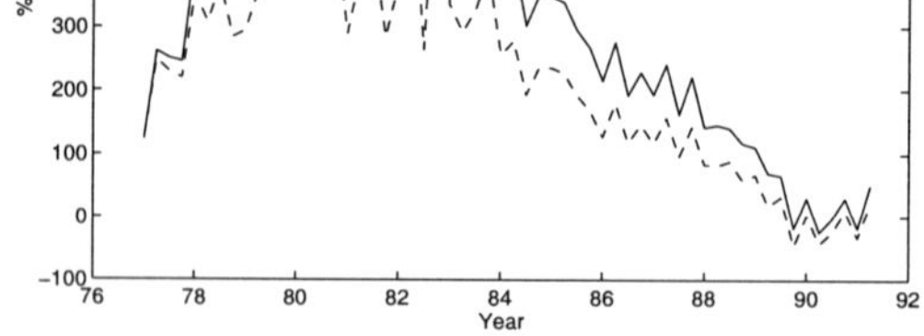

Figure 8.5: Difference Equation Model Control - % Deviations from Base

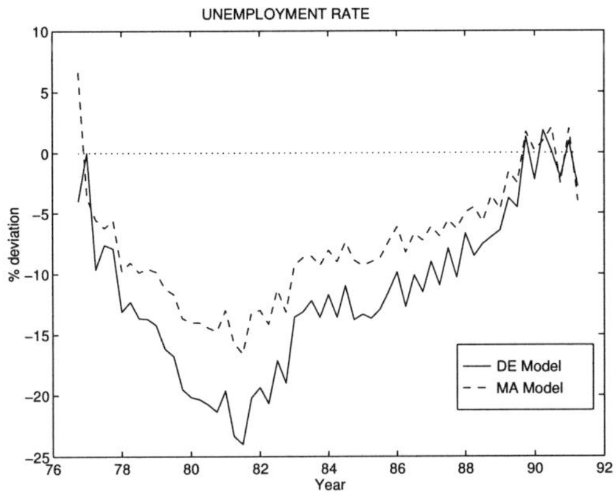

Figure 8.6: Unemployment Rate - % Deviations from Base

ated by the controllers. The plots are in percentage deviation from the baseline results of the nonlinear macroeconometric model. Both controllers generate a fiscal stimulus, which, as can be seen from the uncontrolled trajectory for GNE (Figure 8.2 or 8.3) would be necessary to move the deviations towards zero. The better results of the observer based on the difference equation model (Figure 8.3) are produced by a greater fiscal stimulus (Figure 8.5).

One of the advantages of this type of analysis is that the implications on other key economic indicators can be considered. These indicators are variables that are not the outputs of the model, and thus not included in the objective function, but are, nevertheless, important economic variables. Great care must be taken with interpreting the implications of a counter-factual experiment such as this for so many other variables are assumed to remain the same despite the change in control policies. However Figure 8.6 shows the implications of the modified government expenditure on the Unemployment Rate (URT). The Figure plots the trajectory, for each of the observers with the faster learning gain, as the percentage deviation from base (i.e. no control) for the unemployment rate. For both controllers, the unemployment results are better than with no controller. The higher fiscal stimulus results in the lower unemployment rate. An interesting point from these results is that, from Figure 8.6, the Multiplier Accelerator model moved the Unemployment Rate further from the Historical Baseline than did the Difference Equation model. The opposite is the case for the output of GNE, as can be seen from Figures 8.2 and 8.3.

8.3 Expanded Observers

The first set of experiments used simple single-input, single-output models for the linear model. The second set of experiments uses multiple-input, multiple-output models that reflect current economic theory. The observers also include the more sophisticated control techniques developed in Chapter 4.

The basis model for the observer is the Three State Model developed in Chapter 3 (Section 3.3.5 on page 75). This is a discrete, log linear model based upon IS-LM framework. It was estimated using the same data series and historical data as the nonlinear macroeconometric model; and uses as outputs the same definition of variables as used by the model. The control variables are exogenous to the nonlinear macroeconometric model. They are also defined in the same manner in both models. Thus there are no data linkage problems; and, from a data point of view, the models are in a close neighbourhood.

The model has GNE and the price level as outputs. For the controls it uses the money supply for monetary policy, and the Target Ratio of Public Debt to GDP (RPUBDE) as the fiscal policy control. This latter instrument is effectively a personal income tax instrument for the nonlinear macroeconometric model adjusts the rate of income tax to achieve the target ratio.

The particular model used is that estimated by the state space technique which emphasises the model's ability to track the historical output. The parameters are those of Table 3.5

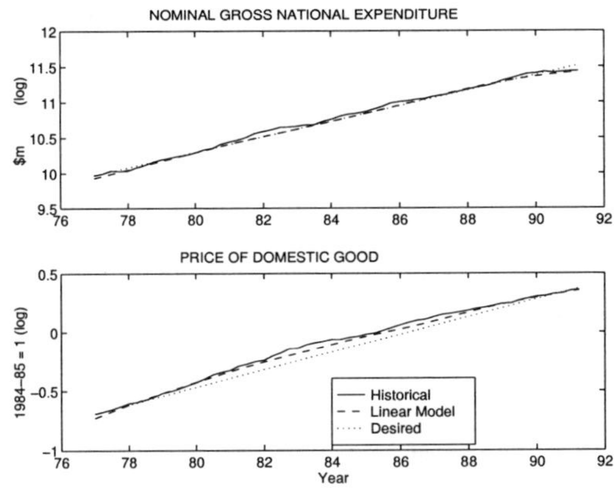

Figure 8.7: Output in Levels

(on page 78), giving the system matrices for the state space form of the model of

$$A = \begin{bmatrix} 0.9911 & -0.0056 & 0 \\ 2.71712 & 0.4893 & -0.8033 \\ 0.1250 & 0 & 0.9077 \end{bmatrix}$$

$$B = \begin{bmatrix} 0.0302 & 0 \\ 0 & -1.9922 \\ 0 & -0.0689 \end{bmatrix} \qquad E = \begin{bmatrix} 0.1302 \\ -10.5231 \\ -0.7318 \end{bmatrix} \tag{8.6}$$

$$C = \begin{bmatrix} 1 & 0 & 0 \\ 0 & 0 & 1 \end{bmatrix} \qquad D = \begin{bmatrix} 0 & 0 \\ 0 & 0 \end{bmatrix}$$

These values give the eigenvalues of A to be $\{0.99765, 0.8862, 0.5253\}$ indicating a slow stable model; the state equation driving matrix is reasonably well-conditioned ($\kappa_2(A) = 19$); and, the rank of both the controllability matrix and the observability matrix for the system is 3, so the system is controllable and observable.

The outputs from the model are shown in Figures 8.7 and 8.8. Figure 8.7 plots the historical levels of GNE and prices, and the estimates from the low order linear model when run out from initial conditions. Also plotted in the Figure are the assumed desired outputs of policy makers. This is the ordinary least squares line through the historical data with the intercept adjusted to be the same as the historical data. This differs from the first set of experiments

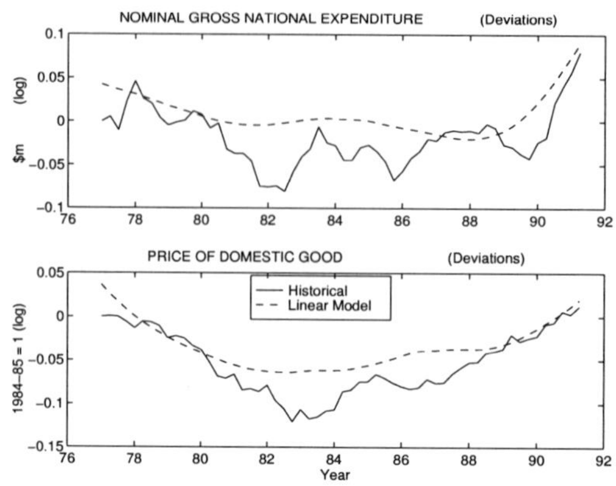

Figure 8.8: Output in Deviations from Desired Output

which used the Historical Baseline for generating desired output. Again this desired output is a set of ramps, and aims at removing the fluctuations in the business cycle from the economy. The ramp price level gives a constant inflation rate of 7.4% p.a. Figure 8.8 plots the model output and historical data as deviations from the desired outputs.

The control technique is the feedforward method of Chapter 4 (Section 4.4 on page 132). The basic approach is to determine the feedforward gain via model inversion, so that after the transients there is no steady state error to a desired output of a set of ramps. The feedforward control is used in conjunction with state feedback control. The state feedback gain matrices are computed by either pole placement or a linear quadratic regulator.

8.3.1 The Experiments

For this set of experiments, four sub-experiments are undertaken. The first, Experiment 2.1, uses pole placement to determine both the stationary state feedback matrix, K, and the stationary learning gain matrix, L. The state feedback matrix is chosen such that the eigenvalues of the closed loop state equation driving matrix, $(A - BK)$ are equally spaced between 0.65 and 0.7; and the learning gain matrix is chosen such that the eigenvalues of the closed loop output-injection driving matrix, $(A - LC)$, are equally spaced between 0.2 and 0.3. In the second experiment, Experiment 2.2, linear quadratic control techniques are used to determine the state feedback matrix. In this Experiment it is assumed that policy makers have stationary weights, and weight each output and control equally but outputs are considered more important and are weighted at 100 times controls. Thus the output deviations weighting matrix, Q, is a 2×2 diagonal matrix with 100 on each diagonal; and, the control deviations weighting matrix, R, is a 2×2 identity matrix. This results in the eigenvalues of the closed loop state equation driving matrix, $(A - BK)$, to be $\{0.76, 0.48 \pm 0.10\imath\}$. The learning gain is chosen in the same manner as in Experiment 2.1.

Experiment 2.3 is the same as Experiment 2.2 except that policy makers increase their weighting of prices ten fold. That is, deviations between actual and desired prices are weighted at 1000 and those for GNE are 100. In Experiment 2.4, it is assumed that the difference between the low order linear model and the Model can be accounted for by known white gaussian noise, and the learning gain is then chosen to minimise this noise. Experiment 2.4 illustrates the use of a Kalman filter. It is assumed that the state equation noise matrix, H (in Equation 4.1 on page 112), is an identity matrix so that independent disturbances effect each equation in the low order model; and that the state noise, $\mathbf{v}$, and output measurement noise, $\mathbf{w}$, are independent and both have covariance matrices of the identity matrix. The state feedback matrix is chosen in the same manner as in Experiment 2.2 making this a linear quadratic gaussian experiment.

8.3.2 Results

The results from the four experiments are presented in Table 8.1. The relative error (as measured by the 2-norm) between the actual output from the nonlinear macroeconometric model, using the particular experiment, and the desired output is used as a metric for the experiment. The table includes the metric for the overall results (i.e. both outputs) and for each individual output separately. For comparison the result with no control applied to the nonlinear macroeconometric model is also presented. This metric is for the Historical

RESULTS				
Experiment		*Relative Error*		
		Both	*GNE*	*Prices*
2.0	Base Model	0.0065	0.0034	0.1800
2.1	Pole Placement	0.0060	0.0023	0.1800
2.2	LQR Equal Weights	0.0062	0.0028	0.1800
2.3	LQR Price Emphasis	0.0063	0.0030	0.1800
2.4	LQG	0.0062	0.0029	0.1800

Table 8.1: Experiment Set Two Results

Baseline results.

Table 8.1 shows that, for each Experiment, the metric for the combined outputs was less than the Historical Baseline indicating that the approach does force the nonlinear macroeconometric model output closer to the output desired by policy makers. It also shows that the gains are greatest in Experiment 2.1 where pole placement is used rather than linear quadratic optimal control. Further, the Table shows that the gains are made in GNE and not in prices. Even when policy makers emphasise the price level by weighting the deviations heavily, the deviations in prices from the desired price trajectory give the same metric. In this case the overall metric is worse than in the case of Experiment 2.2 where there is equal emphasis on outputs. Selecting the learning gain by an optimal minimum variance procedure in Experiment 2.4, did not substantially alter the metric as can be seen from the comparison between Experiment 2.2 and Experiment 2.4.

The metric gives an overall measure of the results, but does not show the temporal differences. Some results are presented visually in Figure 8.9. The GNE results for Experiments 2.1 and 2.2 and for the Historical Baseline are plotted in Figure 8.9. The plots show that GNE is closer to the desired GNE for these experiments.

While the experiments show that the approach leads to GNE being closer to the desired GNE, they also show that there is no change for the price level. An examination of the

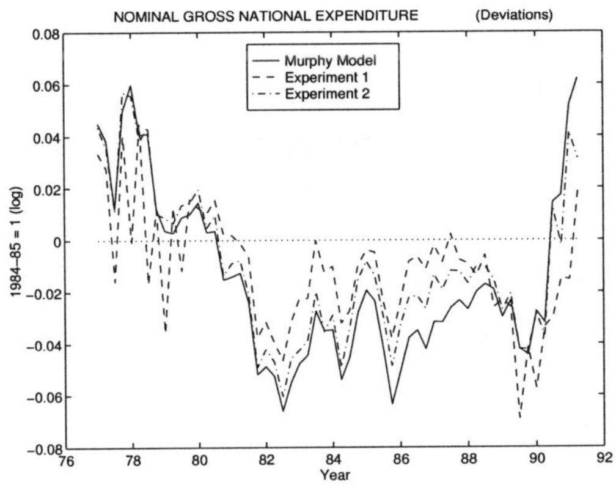

Figure 8.9: GNE Results

nonlinear macroeconometric model indicates the reason for this. The Murphy Model[1] equation for inflation (S12) (Powell and Murphy, 1995, p.129) is basically a lag structure on the growth rate of the medium run equilibrium value, giving the Murphy Model sticky prices. Thus, as far as the price level is concerned, the low order linear model is not a good representation of the nonlinear macroeconometric model. The low order linear model being a good representation of the econometric model is a requirement for the approach to be successful, and these results illustrate this point.

Despite this, it is interesting to note the Money Supply appearing to have no effect on prices in the Murphy Model. One advantage of optimal control exercises like this is that the optimising routine may be useful in finding areas of improvement in the models in that the optimising routine may recommend ridiculous policies.

An important issue is the robustness of the models to parameter uncertainty. This is especially the case here for the low order linear model will not be a true representation of the nonlinear macroeconometric model. Rather than examining the situation with specifically stochastic parameters, as is considered by Amman and Kendrick (1993), the approach

[1] This is the basis model for the dynamic nonlinear macroeconometric model used in the Chapter

is to examine the implications of parameter uncertainty by considering the effects of a perturbation to the parameter matrices. This is the approach used in Chapter 4, and specifically for the feedforward control technique in Section 4.4.2 (on page 134). Here it is extended for the low order model being an observer.

The question of concern here is: what are the effects of parameter uncertainty on the transfer relationship between the inputs to the system and its outputs? The low order deterministic model is

$$\hat{\mathbf{x}}_{t+1} = A\hat{\mathbf{x}}_t + B\mathbf{u}_t + E \tag{8.7}$$

$$\hat{\mathbf{y}}_t = C\hat{\mathbf{x}}_t + D\mathbf{u}_t \tag{8.8}$$

Taking z transforms (Åström and Wittenmark, 1990) and assuming zero initial conditions, then

$$z\hat{\mathbf{X}}(z) = A\hat{\mathbf{X}}(z) + B\mathbf{U}(z) + E \tag{8.9}$$

$$\hat{\mathbf{Y}}(z) = C\hat{\mathbf{X}}(z) + D\mathbf{U}(z) \tag{8.10}$$

So that

$$\hat{\mathbf{Y}}(z) = [C(zI - A)^{-1}B + D]\mathbf{U}(z) + C(zI - A)^{-1}E \tag{8.11}$$

$$= G_o(z)\mathbf{U}(z) + N_o(z) \tag{8.12}$$

where I is the identity matrix.

Thus the dynamics of the relationship between the inputs and outputs of the low order model is determined by the transfer function matrix $G_o(z)$. The condition number of $G_o(z)$, then, indicates the effects of parameter perturbations on this relationship.

Considering the steady state ($z = 1$) and using final values for the ramp input ($\mathbf{r}_{ss}$), then (from Section 4.4.2 on page 134, and as with Equation 4.37) for the deterministic low order linear model in the closed loop

$$\begin{aligned}\hat{\mathbf{y}}_{ss} &= [(C - DK)[W - (A - BK)]^{-1}B + D]\mathbf{r}_{ss} \\ &\quad +(C - DK)[W - (A - BK)]^{-1}E \end{aligned} \tag{8.13}$$

$$= G_c\mathbf{r}_{ss} + N_c \tag{8.14}$$

So $\kappa_2(G_c)$ gives a measure of the effects of parameter uncertainty. If G_c is ill-conditioned, then a small perturbation in the parameters will result in a large difference in the relationship between the inputs ($\mathbf{r}$) of the low order linear model and its outputs ($\hat{\mathbf{y}}$) in the steady state.

Including the output-injection learning to form the observer (of Equation 8.4) then

$$\begin{aligned} \hat{\mathbf{y}}_{ss} &= [N_e(B - LD) + D]\mathbf{r}_{ss} + N_e(E + L\mathbf{y}_{ss}) && (8.15) \\ &= G_e\mathbf{r}_{ss} + N_e(E + L\mathbf{y}_{ss}) && (8.16) \end{aligned}$$

where

$$N_e = (C - DK)[W - (A - BK - LC + LDK)]^{-1} \qquad (8.17)$$

Note that in this situation, the output from the nonlinear macroeconometric model, $\mathbf{y}$ is also a function of $\mathbf{r}$, so that the transfer matrix, G_e, does not give the full relationship.

Finally, the transfer matrices for the entire system which includes the policy makers desired inputs ($\mathbf{y}_{d,ss}$) is given by

$$\begin{aligned} \hat{\mathbf{y}}_{ss} &= G_eF\mathbf{y}_{d,ss} + FME + N_e(E + L\mathbf{y}_{ss}) && (8.18) \\ &= G_s\mathbf{y}_{d,ss} + N_s + N_eL\mathbf{y}_{ss} && (8.19) \end{aligned}$$

Equation 8.19 gives the relationship between policy makers' desired outputs and the outputs of both the observer and the nonlinear macroeconometric model in the steady state. The effects of parameter perturbation can be considered by the condition number of these transfer matrices.

Table 8.2 presents the metric for the robustness of these results to parameter uncertainty. The metric is the condition number of the main transfer matrices, and indicates the magnifying effect of a perturbation to the matrix.

These metrics show that the system is robust to parameter uncertainty. Despite the complex linear equation manipulations the transfer matrices are well-conditioned. They are

ROBUSTNESS OF RESULTS				
		Condition Number		
Experiment		G_c	G_e	G_s
2.1	Pole Placement	123.88	30.28	5.83
2.2	LQR Equal Weights	1.18	2.86	2.42
2.3	LQR Price Emphasis	1.46	2.23	3.06
2.4	LQG	1.18	1.82	1.57

Table 8.2: Experiment Set Two Robustness Metrics

mostly better conditioned than the open loop state equation driving matrix ($\kappa_2(A) = 19$), indicating that the addition of the control improves the robustness. Surprisingly, it is the pole placement method that is the least well-conditioned, but this could be related to the pole placement algorithm. It would be expected that the optimal control and especially the linear quadratic gaussian results to be the least well-conditioned (Doyle et al., 1992). Though, of course, these results are limited to the four Experiments reported here.

The approach here has been to consider parameter uncertainty through robustness of the results. This differs from considering parameter, or other, uncertainty in the devlopment of control policies, as in robust control techniques, and thus whether different policies would be generated in that situation (See, e.g., Amman and Kendrick, 1997).

8.3.3 Policy Implications

Great care must be taken in interpreting the results of counter-factual policy simulations such as those reported here, yet they do create some interesting results. The implications of introducing the policies generated by the approach can be examined by considering Experiment 2.2. This experiment used linear quadratic optimal control and assumes that policy makers emphasise outputs 100 times more than they emphasise controls.

Firstly, consider the control variables. The deviations from the historical values of the control series by those generated in Experiment 2.2 are plotted in Figure 8.10. The Figure plots

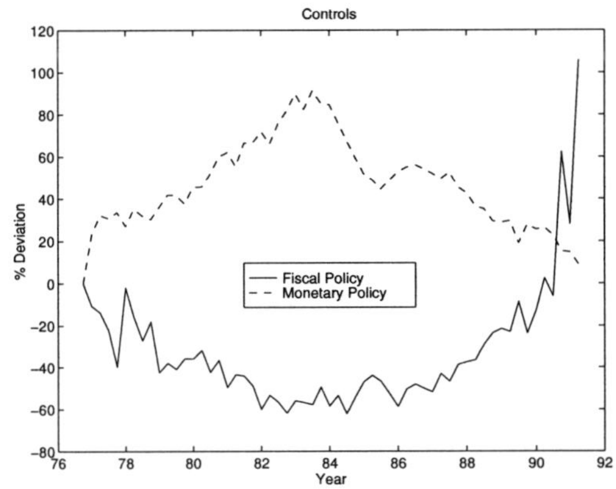

Figure 8.10: Experiment 2.2 % Deviations from Historical Controls

the percentage deviation from its historical value for the fiscal policy control instrument (RPUBDE), and that for the monetary policy instrument (M). The Figure shows that the control policies generated by Experiment 2.2 used substantially greater monetary policy with a mean difference of 46%; and substantially less fiscal policy, with a fall in the mean difference of 33%.

Figures 8.11 and 8.12 show the implications for some other key economic indicators. The Figures plot the temporal percentage difference under Experiment 2 to the Historical Baseline of the nonlinear macroeconometric model. Figure 8.11 plots the percentage deviations for the exchange rate (E) and for long term interest rates (RL). The mean fall in the exchange rate is 13% and the mean fall in long term interest rates is 21%. Figure 8.12 plots government spending (GSPEND) and the unemployment rate (URT). Again both fall relative to their historical values, with the mean for government spending of 2% and a mean of 3% for the unemployment rate.

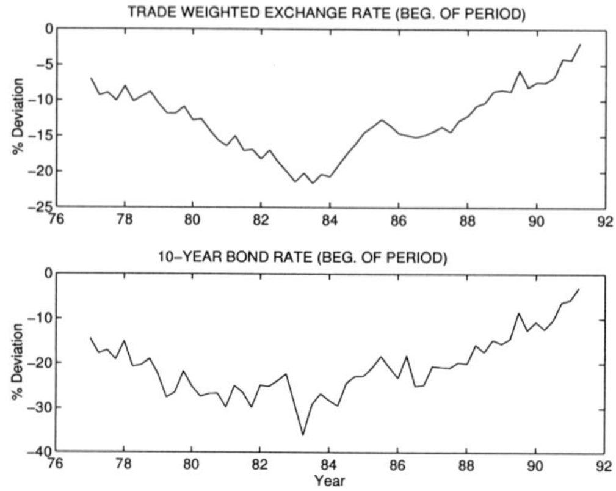

Figure 8.11: Exchange Rate & Interest Rate % Deviations from Base

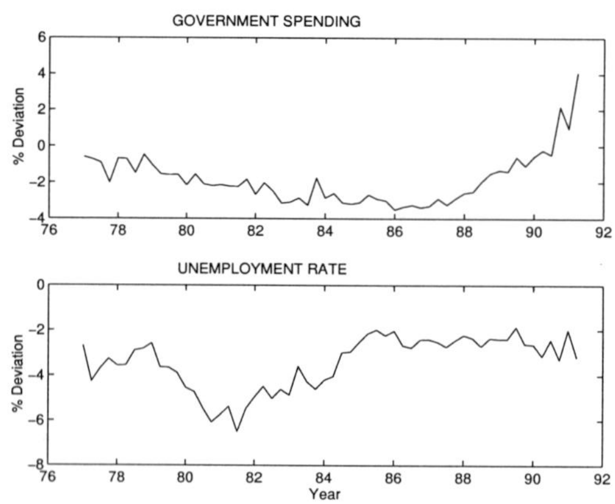

Figure 8.12: Government Spending & Unemployment Rate % Deviations from Base

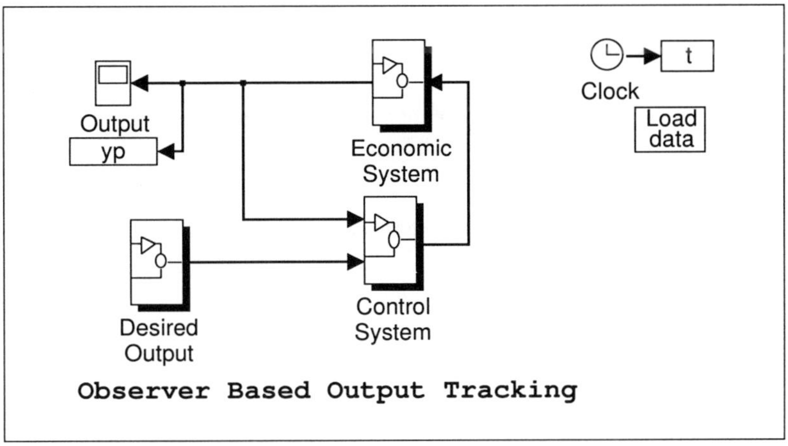

Figure 8.13: Visual Simulation Environment

8.4 Visual Simulation

This Section presents tools that use the low order, linear observer approach with the nonlinear macroeconometric model and that can be used for economic policy generation and analysis through control experiments. The tools are the visual simulation tools considered in Chapter 5, and these allow for a wide range of control experiments to be easily undertaken. The overall method can be seen from the main simulation environment presented in Figure 8.13. The Figure shows the methodology of using a separate control system which is distinct from, but learns from, the economic system for which it is developing control policies. This method compares to the current applications of control theory to economic systems considered in Chapter 6, where the economic system is specifically tailored for control experiments and/or the controls system and the economic system are heavily intertwined.

The economic system used here is the same dynamic nonlinear macroeconometric model used in the earlier experiments in this Chapter. Figure 8.14 shows the visual simulation environment for this model. This differs from that of Chapter 7 (presented in Figure 7.9

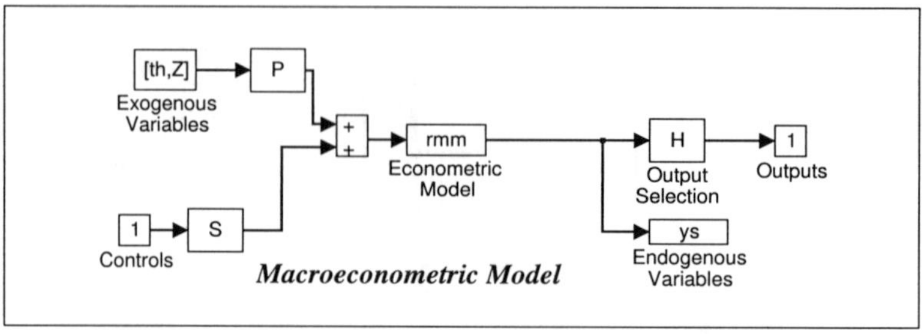

Figure 8.14: Economic System

on page 207), in that the exogenous variables are divided into those completely outside the control experiment and those that are considered as controls. This is the form as presented in Equations 8.3 and 8.4 (on page 213) earlier in this Chapter. The selection matrix objects P and S allow for the incorporation of the controls to the inputs into the macroeconometric model. Similarly, the matrix object H selects the endogenous variables that are to be considered as outputs.

For the experiments reported here to illustrate the visual simulation approach, the linear model used as the basis of the controller is the two input, two output, six state, non-stationary, contemporaneous control model developed in Chapter 4 (in Section 3.3.6 on page 82). The inputs are General Government Purchases of Consumption Goods and the Money Supply, both of which are exogenous to the nonlinear econometric model. The outputs are Nominal Gross National Expenditure and Total Employment. This model is a log linear model with the data normalised to remove the need for the intercepts. This model differs not only in inputs and outputs from the one used in the previous Section, but also in that it is a time series model with contemporaneous control and lags of one year (4 periods).

As is the case with earlier experiments in this Chapter, the experiments use historical data; the linear model uses the same data series and time horizon as the nonlinear econometric model; and, the model was estimated in state space form. The parameters are given in

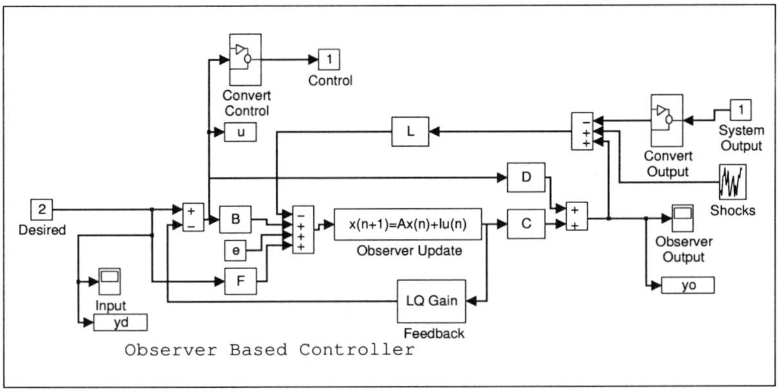

Figure 8.15: Control System

Table 3.16 (on page 104). Again, the linear model is estimated so that it is in a neighbourhood of the nonlinear econometric model. The objective of the control experiments is also to reduce the variability in the business cycle, with the desired output being the straight line (ramp) through the actual historical data. This also helps ensure the models are within a neighbourhood.

An important point with the methodology used here, is that a wide variety of linear models can be used provided they are expressed in state space form. The visual simulation approach allows for easy transfer between different models.

Details of the control system are given in Figure 8.15. This system illustrates the augmented, integral control, ramp tracking approach that was developed in Chapter 4 (Section 4.3 on page 114), whereby, in this case, the basis model for the observer is augmented by an additional four states. With the visual simulation, the feedback gains of the linear quadratic regulator can be easily modified so that the trade offs between various degrees of emphasis on outputs and controls can be examined. So, too, can other components such as the power of the shocks. Further, intermediate results, such as the observer output, can be examined during the simulation and are stored for later use.

As the nonlinear econometric model data is in levels and the observer data is in logarithms,

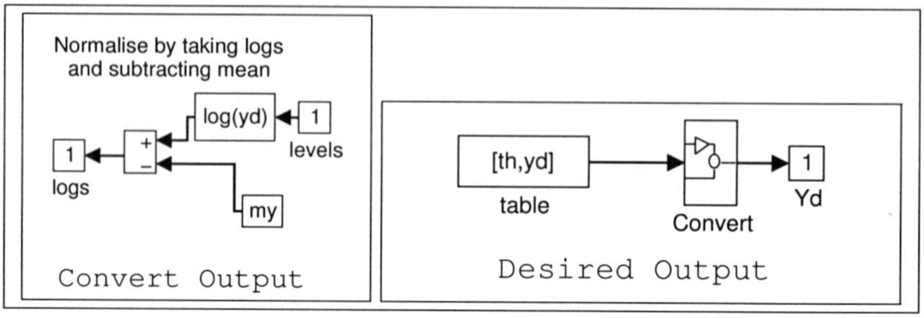

Figure 8.16: Output Conversion and Desired Output

RELATIVE ERROR *Between Actual & Desired Output*			
Weight	GNE	Employment	Both
0	0.106591	0.044003	0.106090
1	0.096876	0.035179	0.096391
10	0.092479	0.034353	0.092016
50	0.088002	0.034786	0.087574
100	0.087879	0.034821	0.087452
500	0.086931	0.034212	0.086507
1000	0.089872	0.034330	0.089430
2000	0.091814	0.034380	0.091359

Table 8.3: Summary Metrics

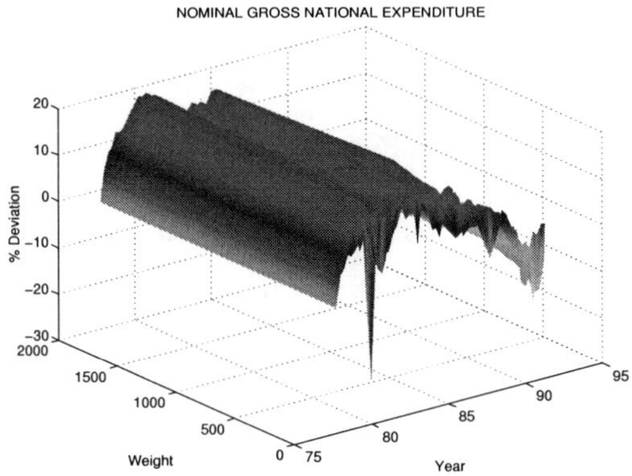

Figure 8.17: GNE: % Deviations from Desired Output

the data is converted as part of the control system. Figure 8.16 presents the details of the conversion of the levels data. This conversion is also used in the calculation of the desired output. The visual simulation for calculating desired output is given in Figure 8.16, and it shows how a look-up table is used to generate the actual values. The look-up table allows for both interpolation and extrapolation of values.

The results of a set of experiments undertaken with this visual simulation environment are presented in Figures 8.17 and 8.18. The experiments involve a variation in the weights of the linear quadratic regulator. The Figures plot the percentage deviation from the desired output for each of the outputs of Gross National Expenditure and Total Employment. For these experiments both outputs were equally weighted, and both controls were equally weighted, so that it is concerned with the trade-off between outputs and controls rather than, say, between outputs. The Historical Baseline result, which is that generated without the controller, is presented as having zero weights. The Figures show that, like the earlier experiments in this Chapter, the approach generates control policies whereby the nonlinear econometric model is closer to the desired output, than is the case when the controls take their historical values. This can also be seen in Table 8.3 where the summary metrics of the relative error (using the 2-norm) between the actual and desired output for Gross National

Figure 8.18: Employment: % Deviations from Desired Output

Expenditure, for Total Employment, and for the combined outputs are presented. Again, the zero weighting indicates the Historical Baseline results.

These experiments illustrate the use of the visual simulation tools. As with the earlier experiments, it is possible to consider the implications of policies on other key economic indicators. These experiments have focused on historical data, but if the exogenous variables are predictions of the future trajectories of these variables, then the visual simulation tools can be used in policy analysis with these predictions. This is in a similar manner to which they were used with the steady state solution of the nonlinear econometric model.

The power of the visual simulation tools lie in the ease with which policy analysis through control experiments can be undertaken. They quickly generate results, aid in the interpretation of these results, and visually show the details of the experiments.

8.5 Conclusion

This Chapter has synthesised much of the earlier work in this book. It has shown how separate model based control can be used in policy development and evaluation of a large practical econometric model using actual data. The lower order observer based technique has been illustrated with a number of the linear models built in earlier Chapters to develop control policies for the large nonlinear econometric model. The observer's output-injection learning helps to compensate for the differences in the models.

One of the advantages of the approach is that the control system is not embedded in the system it is controlling, and hence enhancements to the control system can be made without reference to the economic system. This has allowed the observer to be modified so that it has the ability to track outputs of a set of ramps. Such desired outputs are important where the aim of policies is to reduce the business cycle around the trend in economic growth.

The Chapter has presented the results from a large number of experiments using the approach with models built in the book. These experiments show that the approach has merit. In particular, it is possible to develop control policies for the large nonlinear econometric model on the low order linear models that reduce the variability in the business cycle.

To the control engineer the results may not be spectacular for the process is not rapidly brought into line with the desired output. Yet the economy is a slow plant with large time constants, complex and intricate feedback loops, high costs to control effort, and is not as well understood or modelled as many physical processes. Policy makers may be sceptical of rapidly converging zero error tracking results as this differs from their experience. This is why many and varied experiments are important. Care must be taken that the techniques are not rejected out of hand.

The latter part of the Chapter was concerned with tools that aid in the use of the approach in policy development and evaluation. Most previous studies simply present a set of results. But the visual simulation tools allow for many experiments to be undertaken and the results to be easily interpreted. Thus 'what if' types of questions can be easily considered, and the trade off between various policy approaches and objectives can be evaluated. For practical use by policy makers, this is a desirable property. Further, the visual simulation approach gives the ability to swap between different models, so that they, too, may be compared.

For the presentation of the methodology and the ease of experimentation, these visual simulation tools are beneficial.

Chapter 9

NONLINEAR OBSERVERS

9.1 Introduction

The purpose of this Chapter is to develop nonlinear observers for use with economic systems. Previous Chapters have focused on the use of linear observers for linear systems, or on linear observers for nonlinear systems. It would seem reasonable that a nonlinear observer would perform better than a linear observer in learning the states and developing controls for a nonlinear system.

The problem is that the theory and application of nonlinear observers is not well developed. This compares to the theory and application of linear observers. This Chapter extends the theory of linear observers into nonlinear observers and, in doing so, is able to use the well developed theory of linear observers.

Nonlinear observers are the focus of much current research in the control engineering literature (e.g. Bestle and Zeitz, 1983; Krener and Isidori, 1983; Walcott, Corless and Zak, 1987; Lin and Byrnes, 1995; Eker and Åström, 1996). Some of this research (e.g. Moraal and Grizzle, 1995) focuses on the issue of learning the states of the system, given a model, so that the issue of using these states to generate a set of controls is not addressed. For the observers developed here, both these issues are addressed.

Given that the model of the system is known, then there are two major issues with the use of observers. The first is to consider if it is possible for the observer to learn the unknown values of some or all of the system's states. These unknown states may result from disturbances which may be in the initial value of the states, as is the usual control engineering use of observers, or may result from exogenous disturbances. The second major issue is to consider that, once the states are learnt, whether these learnt states can be used to develop control policies for the system. Another important problem is that, if the model of the system is not be known with certainty, then what are the implications for the two previous issues.

Chapter 2 of this book was concerned with the use of an observer for learning the expectations of forward looking economic actors. It showed that, with linear models, the observer could be used to learn these (and other) states. The learnt states could then be used with conventional linear control techniques (including linear quadratic optimal control) to develop control policies for such models. This Chapter returns to the issue of learning the forward looking expectations of economic actors, but for nonlinear models. These forward looking expectations may be states in the system, and solved using the rational expectations hypothesis. The problem is, then, the nonlinear analogue of that considered in Chapter 2, that the initial values of these states are unknown. This Chapter develops nonlinear observers that are able to learn about these unknown initial initial values.

The rational expectations hypothesis is a particular approach to forward looking expectations. If a a broader view of expectations is used, so that these variables are simply variables that 'jump' to 'news', than they may be treated as disturbances that shock the system. This Chapter also considers forward looking expectations in nonlinear models in this perspective.

The Chapter continues the theme of using a separate control model for the development of control policies, and then applying this to the system of interest. This is the basis of the observer based approach. Yet in this book, the control model has been known to be different from the system for which it is generating control policies. In particular, the control model has been of a lower order than the economic system. In this Chapter, in a direct extension of Chapter 8, a nonlinear observer model is used to generate policies for the high order nonlinear econometric model developed in Chapter 7.

9.2 Preliminaries

The development of the observers will be presented in continuous time. This is for convenience, as with both the techniques developed in this Chapter, the discrete time case is analogous to the continuous time case. Let the system be described by the following nonlinear model:

$$\dot{\mathbf{x}}(t) = \mathbf{h}(\mathbf{x}(t), \mathbf{u}(t), \mathbf{z}(t), t) \quad (9.1)$$

$$\mathbf{y}(t) = C\mathbf{x}(t) \quad (9.2)$$

Here, Equation 9.1 is the state equation and Equation 9.2 is the output equation. In the system $\mathbf{x} \in \Re^n$ is a vector of states, $\mathbf{u} \in \Re^p$ is a vector of controls, $\mathbf{z} \in \Re^p$ is a vector of exogenous variables and $\mathbf{h}(.)$ is a set of n equations. In the output equation, $\mathbf{y} \in \Re^m$ is a vector of outputs, and $C \in \Re^{m \times n}$ is the output selection parameter matrix. It is assumed that the outputs are a linear combination of the states. This is the usual case in economic models. The states may include forward looking expectations variables, with unknown initial values. Here it is assumed that the initial values of all states are unknown. From the point of view of the model, the exogenous variables are known for all the time horizon.

For convenience, and without loss of generality, the model will be simplified further for developing the observers. Firstly, let the system be time invariant. Secondly, the essence of the dynamics of the system, in the closed loop, is determined by the states and the controls, so that the exogenous variables are not considered in the development of the observers. In the illustrative application of the techniques, in Section 9.4, the exogenous variables are included.

Thus for the development of the observers, concern is with economic systems given by the state and output equations

$$\dot{\mathbf{x}}(t) = \mathbf{f}(\mathbf{x}(t), \mathbf{u}(t)) \quad (9.3)$$

$$\mathbf{y}(t) = C\mathbf{x}(t) \quad (9.4)$$

It is assumed that this model is known, but the problem is to generate control policies for the model given that the initial values of the states are unknown.

The focus here is on closed loop systems, with the controls generated on the observer's states. The economic system is the model given by Equations 9.3 and 9.4. It has initial values for its states, say ξ, but they are unknown. The observer's states, which will be, again, indicated by $\hat{\mathbf{x}} \in \Re^n$, have known (allocated) initial states of $\hat{\xi}$. The control rules are built using the states from the observer. Again, for simplicity, but without loss of generality, these will be of the form

$$\mathbf{u}(t) = -K(t)\hat{\mathbf{x}}(t) \tag{9.5}$$

where $K(t) \in \Re^{p \times n}$ are the state feedback gain matrices. For some observers, the feedback matrix be considered as time invariant and indicated as K. Time invariant feedback gains have the advantage of producing time-consistent feedback rules (Currie and Levine, 1993*b*).

9.3 A Nonlinear Observer

A standard technique (e.g. Bryson and Ho, 1975) with such models is to linearise the model around some reference point or some reference trajectory. The former case produces a time invariant system, and the latter case produces a time varying system. In both cases, the linear model can be considered to be in deviations around the reference. There is a truncation error associated with the linearisation, so that the model is only applicable in the neighbourhood of the reference.

The linearised model is then used with linear techniques. If an observer is to be generated to learn the states, then it is formed on the basis of the linear model. But given that the nonlinear model is known, then it may be used as the basis model for the observer instead of the linear model. The observer can be formed by using output-injection learning, with the outputs being injected into the nonlinear model. This produces an observer of the form

$$\dot{\hat{\mathbf{x}}}(t) = \mathbf{f}(\hat{\mathbf{x}}(t), \mathbf{u}(t)) + L(t)(\mathbf{y}(t) - \hat{\mathbf{y}}(t)) \tag{9.6}$$

$$\hat{\mathbf{y}}(t) = C\hat{\mathbf{x}}(t) \tag{9.7}$$

This observer is of the same type as the observer introduced in Chapter 2 (Equation 2.22 on page 30), but uses the nonlinear model. This basis model is extended with the addition of the learning term. Again it is the purpose of this term to learn the differences between the observer and the system. The learning gain matrices, $L(t) \in \Re^{n \times m}$, determine the emphasis of the term. In Equation 9.6 these are shown to be time varying.

The immediate question is in the selection method of the learning gain. The technique proposed here is that it be determined in the usual linear manner using the linearised model. That is, rather than using the linearised model to form the basis model in the observer, it is used to determine the matrix gains to be used in conjunction with the nonlinear model.

Linearisation around a reference point generates a time invariant linear model, so the associated learning gain matrix may be time invariant. If the linearisation is around a time varying reference trajectory, then the linear model will be time varying. This reference trajectory is usually some known and meaningful trajectory associated with the control problem. In this case, the associated learning gain matrix will be time varying. These are the standard approaches. A further approach is proposed here. This is that the linearisation occur within the closed loop. In this case, the reference trajectory is taken to be the observer's trajectory. The aim is to have the system and the observer in as close a neighbourhood as possible. This linearisation may be periodic, so that there is effectively a sliding window in time of the use of each linearised model. The resultant observer is a nonlinear periodically time varying observer with the learning gain matrix computed at each linearisation. The learning gain matrix is computed after a linearisation of the model at the beginning of each time window.

A similar approach can be used for determining the state feedback gain matrix. The matrix can be generated with the linearised model in the conventional manner using linear theory. If the linearisation is around a reference trajectory resulting in a time varying linear model, then the state feedback gains will be time varying. A time invariant gain matrix could be used with a single linearisation of the model. Further, the periodic linearisation in the closed loop could be used to generate a periodically time varying state feedback gain matrix. Standard linear techniques could be used with the gain matrix computed by pole placement or a linear quadratic regulator. More complicated linear techniques, such as the linear quadratic tracking or feedforward techniques of Chapter 4, could also be used.

The important question is: how applicable is the linear theory and the gain matrices determined from the linearised version of the system to the nonlinear model?

Consider the case of linearising the state equation (Equation 9.3) around some reference trajectory $(\mathbf{x}_0(t), \mathbf{u}_0(t))$. Define

$$\delta\mathbf{x}(\mathbf{t}) = \mathbf{x}(\mathbf{t}) - \mathbf{x}_0(t), \qquad \delta\mathbf{u}(\mathbf{t}) = \mathbf{u}(\mathbf{t}) - \mathbf{u}_0(t), \tag{9.8}$$

then, by Taylor's Theorem,

$$\delta\dot{\mathbf{x}}(t) = \frac{\partial\mathbf{f}(\mathbf{x}(t), \mathbf{u}(t))}{\partial\mathbf{x}(t)}\delta\mathbf{x}(t) + \frac{\partial\mathbf{f}(\mathbf{x}(t), \mathbf{u}(t))}{\partial\mathbf{u}(t)}\delta\mathbf{u}(t) + O((\delta\mathbf{x}(t))^2, (\delta\mathbf{u}(t))^2) \tag{9.9}$$

In Equation 9.9 the partial derivatives are evaluated along the reference trajectory, $(\mathbf{x}_0(t), \mathbf{u}_0(t))$; and $O(r)$ refers to order r. The last term in the Equation is a remainder term, and it incorporates the higher order terms of the Taylor Series. It is a function which is principally determined by the size of the neighbourhood of the system from the reference trajectory.

Equation 9.9 can be written as

$$\dot{\mathbf{x}}(t) = A(t)\mathbf{x}(t) + B(t)\mathbf{u}(t) + E(t) \tag{9.10}$$

In Equation 9.10 the states and controls are in deviations around the reference trajectory. The remainder term ($E(t)$) can be considered as another forcing function measuring the implications of truncating the higher order terms. The effects of this forcing function decrease quadratically the closer the system is to the reference trajectory.

Using this linearised state equation generates the linearised version of the model (Equations 9.3 and 9.4) given by

$$\dot{\mathbf{x}}(t) = A(t)\mathbf{x}(t) + B(t)\mathbf{u}(t) + E(t) \tag{9.11}$$

$$\mathbf{y}(t) = C\mathbf{x}(t) \tag{9.12}$$

The linear model is of a similar form to the linear models used in earlier Chapters. The difference is that it is a time varying model, but this makes little difference to many of the techniques used in those Chapters. The linear model differs from the usual control engineering state form in that it includes the extra term, the truncation error forcing term, $E(t)$. In the model, the state equation driving matrices, $A(t)$, and the control forcing matrices, $B(t)$, are the respective partial derivatives. The derivatives are evaluated at, and thus are dependent upon, the reference trajectory. Considering this dependence as introducing uncertainty into these matrices, and considering the truncation error as exogenous, allows the linear state space form to be of the same type as the linear models used in earlier Chapters.

With this linearised model the observer learning gain and the state feedback gain matrices can be computed using linear techniques, in the same manner as in earlier Chapters, if the linear model has the appropriate properties, such as controllability and observability. Thus $L(t)$ in Equation 9.6, and $K(t)$ in Equation 9.5 are determined on the basis of Equations 9.11 and 9.12.

The linear theory behind the observer was considered in Chapter 2 (in Section 2.4 on page 26). The question arises as to what are the implications of the linearisation on this theory. Of particular interest is the computation of the control gain matrices, the computation of the learning gain matrices, and the separation between the two sets of matrices. For convenience, and without loss of generality, these issues will be examined using a time invariant linear model, and time invariant gain matrices.

Firstly consider the effects of linearisation on the learning error. The learning error is the difference between the system states and the observer states, so that $\tilde{\mathbf{x}}(t) = \mathbf{x}(t) - \hat{\mathbf{x}}(t)$. The growth in this error is given by

$$\begin{aligned}
\dot{\tilde{\mathbf{x}}}(t) &= \dot{\mathbf{x}}(t) - \dot{\hat{\mathbf{x}}}(t) \\
&= \mathbf{f}(\mathbf{x}(t), \mathbf{u}(t)) - \mathbf{f}(\hat{\mathbf{x}}(t), \mathbf{u}(t)) - L(\mathbf{y}(t) - \hat{\mathbf{y}}(t)) \\
&= A\mathbf{x}(t) + B\mathbf{u}(t) - A\hat{\mathbf{x}}(t) - B\mathbf{u}(t) - L(\mathbf{y}(t) - \hat{\mathbf{y}}(t)) + \check{E}(t) \\
&= (A - LC)(\mathbf{x}(t) - \hat{\mathbf{x}}(t)) + \check{E}(t)
\end{aligned} \tag{9.13}$$

where

$$\breve{E}(t) = O((\delta\mathbf{x}(t))^2, (\delta\hat{\mathbf{x}}(t)^2), (\delta\mathbf{u}(t))^2) \tag{9.14}$$

In Equation 9.13 both the observer and the system (or process) are linearised around the same reference point. The growth in the learning error is determined by both the eigenvalues of the driving matrix, $(A - LC)$, and the truncation error forcing function. If the effects of the truncation error forcing function are small relative to the driving matrix, then the learning gain may be chosen to drive the learning error towards zero. The observer may then learn the states adequately.

The effects of the truncation error forcing function term are quadratically dependent upon the 'closeness' of the system to the reference point. A metric for these effects is given by the distance the system, observer and controls are away from the reference point. This metric also gives a measure of the uncertainty in the linear state equation driving matrix, A, and hence the learning gain, L, chosen on the basis of it. The is because A, being the partial derivatives evaluated at the linearisation point, is dependent upon this linearisation point.

The effects of using the linearised state equation on the relationship between the selection of both sets of gains by linear methods, can be seen by considering the closed loop system. Substituting the control rule (Equation 9.5) into the linearised state equation (Equation 9.10) gives

$$\begin{aligned}\dot{\mathbf{x}}(t) &= A\mathbf{x}(t) + B\mathbf{u}(t) + E(t) \\ &= A\mathbf{x}(t) - BK\hat{\mathbf{x}}(t) + \breve{E}(t) \\ &= [A - BK]\mathbf{x}(t) + BK\tilde{\mathbf{x}}(t) + \breve{E}(t)\end{aligned} \tag{9.15}$$

Combining Equations 9.13 and 9.15, gives

$$\begin{bmatrix}\dot{\mathbf{x}}(t) \\ \dot{\tilde{\mathbf{x}}}(t)\end{bmatrix} = \begin{bmatrix}A - BK & BK \\ 0 & A - LC\end{bmatrix}\begin{bmatrix}\mathbf{x}(t) \\ \tilde{\mathbf{x}}(t)\end{bmatrix} + \begin{bmatrix}\breve{E}(t) \\ \breve{E}(t)\end{bmatrix} \tag{9.16}$$

where 0 is the $n \times n$ matrix of zeros. As the driving matrix in Equation 9.16 is block triangular, then its eigenvalues are the eigenvalues of the blocks. Thus the selection of L and K are separate. This is analogous to the results of Chapter 2 (on page 31). But notice that, in this linearised case, there is the forcing function resulting from the truncation error. While this will not alter the result that the eigenvalues determined by the selection of the learning gains and the state feedback gains are separate, it does mean that the gain selection is only theoretically applicable in a neighbourhood of the reference point. The linear system matrices are dependent upon the reference point.

One way of examining the effects of the truncation forcing function is to consider that the linear system is uncertain, as it is dependent upon the reference point. A measure for this uncertainty is to consider the implications of perturbations on the eigenvalues of the driving matrix, $(A - LC)$, in the learning error equation (Equation 9.13). Similarly, the effects of perturbation on the eigenvalues of the closed loop state equation driving matrix, $(A - BK)$, can be used to measure this uncertainty. Again, the condition numbers of the associated eigenvector matrices are an appropriate metric.

In Chapter 8, the economic system was a nonlinear macroeconometric model. In that Chapter the observers used linear models, and they were of lower order than the macroeconometric model. It was a similar argument to that above that justified their use.

Using the nonlinear, rather than linearised, process for the basis model in the observer means that, *ceteris paribus*, there is less strength in the truncation error forcing function. The reduced strength is because there is only truncation error in the selection of the learning and feedback gains. While these gains are a powerful determinant of the closed loop dynamics, the reduction in any truncation error will be beneficial. Given that the observer learns the states and generates a control that stabilises the system, then repeated linearisations of the closed loop may improve the performance of the observer. This is because it may reduce the effects of the truncation error forcing function and the uncertainty in the system matrices at each linearisation. As the system states, observer states and controls approach their steady state values, repeated linearisations around, say, the observer position, will reduce the distance between the reference point and the system, and the reference point and the observer. This will continually reduce the effects of the truncation error forcing function.

The cost of repeated linearisations is the increased computational overhead. These computational overheads can be reduced by periodic linearisations. An approach is to have a window of time for each linearisation, and thus each learning and feedback gain matrix, while the strength of the truncation error forcing function is above some predetermined limit. A metric of the truncation error is the distance of the observer and system from the reference point. While the metric is large, a linearisation is done at each time step. That is, linearise whenever the truncation error is greater than some prescribed tolerance.

The truncation error can also be used as a metric as to the appropriateness of this nonlinear observer technique. If the truncation error is large, for the linearisation at a given reference point, then the technique may not be appropriate. If this is the case for a number of reference points, then the technique may be inappropriate.

A metric can be given by

$$\epsilon(t) = \max\{||\mathbf{x}(t) - \mathbf{x_0}(t)||_2, ||\hat{\mathbf{x}}(t) - \mathbf{x_0}(t)||_2, ||\mathbf{u}(t) - \mathbf{u_0}(t)||_2\} \tag{9.17}$$

Now $\epsilon(t)$ can be computed at each time step without the need for linearisation, and it can be used as a metric of the 'closeness' of the linearised model to the nonlinear system. It then indicates the strength of the truncation error forcing function, and the uncertainty in the linear model system matrices and resulting gain matrices. From it, the following algorithm can be used. While $\tau_l \leq \epsilon(t) \leq \tau_m$ use the current linearised model. If $\epsilon(t) < \tau_l$, then recompute the linear model and compute gain matrices. If $\epsilon(t) > \tau_m$, then the nonlinear observer technique should not be used.

The advantage of periodic linearisation is that it can be used to overcome the computational overhead associated with each linearisation. While computational speed may be important in time critical systems, it is generally not a problem with the type of economic models used here.

9.4 Application to a Low Order Model

The nonlinear observer is illustrated with a low order nonlinear model constructed in the IS-LM framework. A suite of IS-LM type models was developed in Chapter 3 (Section 3.3

on page 67), but there the emphasis was on linear models. This nonlinear model has two states of national income and interest rates. These are also considered as the outputs. The model has two controls of a fiscal policy instrument and a monetary policy instrument. It also contains exogenous variables. This nonlinear model is, then, of the form given by Equations 9.1 and 9.2.

The model is

$$\frac{dy(t)}{dt} = \alpha_1 y(t) - \alpha_2 y(t)r(t) + \alpha_3 f(t) + \alpha_4 y^e(t) \tag{9.18}$$

$$\frac{dr(t)}{dt} = -\beta_1 r(t) + \beta_2 y(t)r(t) - \beta_3 m(t) + \beta_4 r^e(t) \tag{9.19}$$

The variables are: y is national income (say, Gross Domestic Product); r is interest rates; f and m are fiscal and monetary policy controls, respectively; and, y^e and r^e are the exogenous expectations of forward looking economic actors for national income and interest rates respectively. Equally well, other exogenous variables could be included in this framework. The time invariant parameters of the model are indicated by Greek letters.

This model is of the IS-LM framework with Equation 9.18 being an IS function and Equation 9.19 being an inverted money demand function. It is a nonlinear, adjustment process form model, and can be considered as explaining the movement of the economy back towards it long run equilibrium. The adjustment process form models were developed in Section 3.3.2 (on page 69).

In this model, expectations of forward looking economic actors are considered to be exogenous, in that they are determined outside the model. Thus the rational expectations hypothesis is not being used. These variables are free to 'jump' to 'news' from outside the model. Here they are treated more as disturbances to the model. It is the aim of the observer to learn the effects of these variables.

In this model, the fiscal and monetary control effects are linear. So, too, are the the effects from the expectations of forward looking economic actors. The nonlinearities enter the model through the relationship between national income and interest rates. This relationship is of the Lotka-Volterra equation type, making the model an extended predator-prey

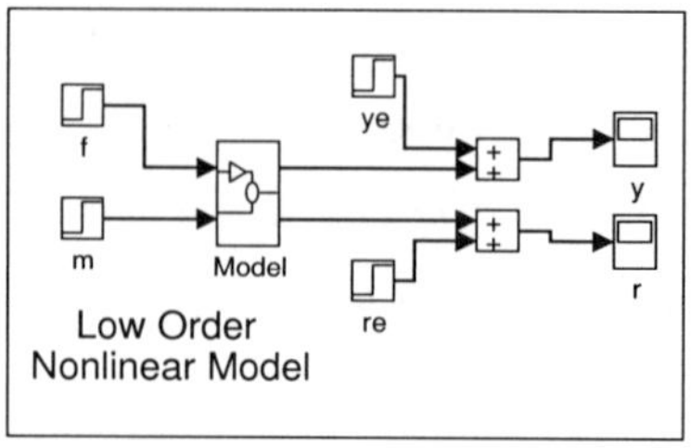

Figure 9.1: Simulation Environment for Low Order Nonlinear Model

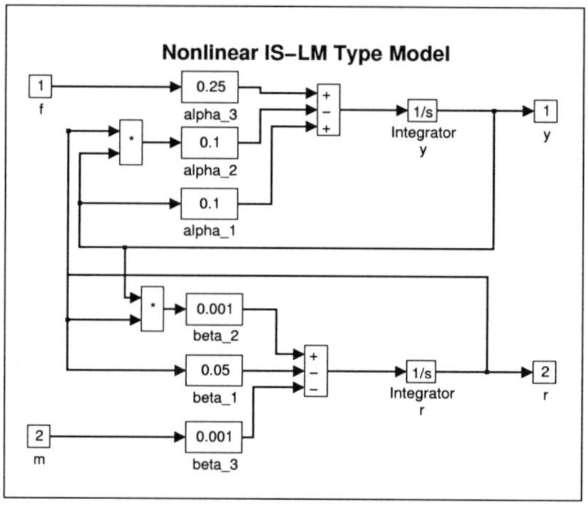

Figure 9.2: Low Order Nonlinear Model

type model. Lotka-Volterra models have been used previously in economics in, for example, Goodwin's (1990) model of the class struggle.

Defining y and r as the states and outputs; f and m as the controls; and, the expectations variables as exogenous disturbances; then the model can be written as

$$\dot{\mathbf{x}}(t) = \mathbf{f}(\mathbf{x}(t), \mathbf{u}(t)) + G\mathbf{x}^{\mathbf{e}}(t) \tag{9.20}$$

The visual simulation environment for the model is presented in Figure 9.1. It is a simulation environment of the type considered in Chapter 5. The simulation environment allows for the simulation of the step response to inputs for the model, the initial condition response, and the effects of a step in the expectations variables. For clarity, the simulation assumes the value of the expectations parameters (α_4 and β_4) are determined by the user in the step size (or are unity). The outputs of the model are plotted during the simulation. In the Figure, information hiding is used so that the details of the model are hidden in the *Model* object.

Figure 9.2 presents details of the *Model* object. This is the visual simulation of Equations 9.18 and 9.19 excluding the exogenous variables. Figure 9.2 visually presents the flow of data through the model. The parameter values are generated by modifiable objects. Selecting such objects (e.g. *beta_1*), opens a slider-bar window where the value of the parameter can be easily altered. This may occur interactively, so the user can readily see the effects of different parameter values on the model. This is particularly useful for this model, which is highly sensitive to parameter values.

For the following simulations, the parameter values are those shown in Figure 9.2. That is, $\alpha_1 = 0.1$, $\alpha_2 = 0.1$, $\alpha_3 = 0.25$, $\beta_1 = 0.001$, $\beta_2 = 0.001$, and $\beta_3 = 0.05$. The forward looking expectations parameters are $\alpha_4 = 1$ and $\beta_4 = 1$. The initial conditions are $y(0) = 100$ and $r(0) = 1$.

The initial condition response of the open loop model is presented in Figure 9.3. For the simulation, the exogenous variables and controls take their long run equilibrium values of zero. It can be seen from the Figure, that the model generates a good nonlinear response for the illustration of the observer techniques. It generates an oscillatory business cycle in deviations in national income and interest rates from their long run equilibrium. The Figure shows the typical response of the Lotka-Volterra system with an oscillatory pattern in the output trajectories for the time horizon.

This economic model is highly nonlinear. It thus makes an ideal test model for the nonlinear observer techniques developed above. The model is very sensitive to its parameters, and this adds to its attractiveness as a test model.

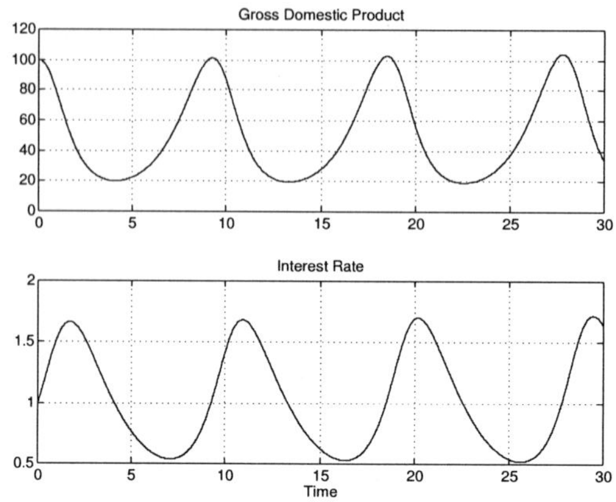

Figure 9.3: Nonlinear Model: Open Loop

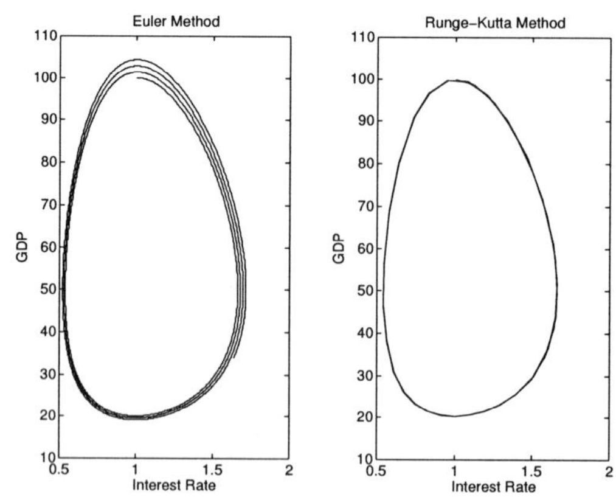

Figure 9.4: Nonlinear Model: Phase Plane

The simulation can also be used to highlight the effects of truncation error. The simulation in Figure 9.3 uses an Euler solution method (Burden and Faires, 1985) for the differential equation (Equation 9.20). This method has similarities to the nonlinear observer technique in that it uses a first order Taylor Series expansion. It thus has a local, or each time step, truncation error of the order of the time step size squared. This is the same result as for the nonlinear observer, except the step size is in the states and controls from the reference point. The accumulation of this truncation error in the solution of the differential equation highlights the importance of truncation error.

This simulation is of the Lotka-Volterra components of the model, so it would be expected that the repeating pattern would continue for all time. A close examination of Figure 9.3 shows that this is not the case. This can be more easily seen from Figure 9.4 which plots the phase plane of the two outputs. The Figure plots the phase plane for the simulation using the Euler and Runge-Kutta 4/5 solution methods. The Runge-Kutta method has a much lower truncation error than the Euler method (Burden and Faires, 1985). The Figure shows the Euler method does not produce the repeatable pattern. It can be seen from Figure 9.3 that the response is gradually exploding. It is the accumulation of truncation error that is the cause. Figure 9.4 shows that the Runge-Kutta method does generate the repeatable pattern for the effects of truncation error is less.

The example shows the strength of the truncation error forcing function. The truncation error in this example has been in the solution method in the time step, so in this case, the truncation error accumulates in the solution. This truncation error can, eventually, swamp the dynamics of the model. This was shown in more detail in Chapter 5 (in Section 5.4 on page 167). Truncation error in the state and control space, as in the case of the nonlinear observer, can, analogously, also have a powerful effects.

It is the objective of the nonlinear observer to learn the states of the model, and to develop control policies for the model. As this model is very sensitive to parameter values, and disturbances, this makes these objectives for the observer a difficult task.

The economic system, or process, is the nonlinear model with parameters and initial con-

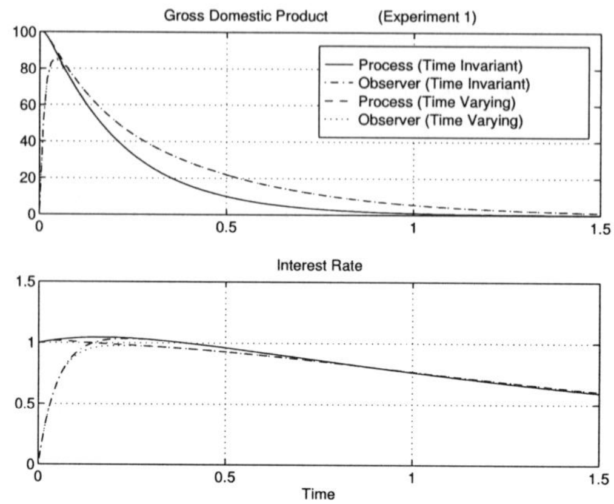

Figure 9.5: Nonlinear Observer: Experiment 1

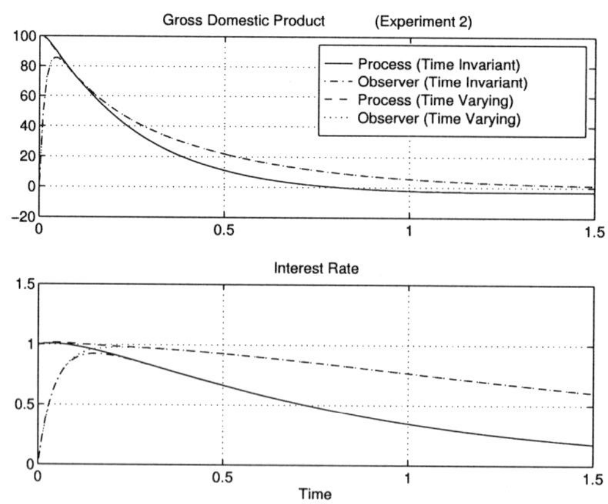

Figure 9.6: Nonlinear Observer: Experiment 2

ditions of those used above. The process is given by Equation 9.20, with the outputs being both the states. The nonlinear observer is given by

$$\dot{\hat{\mathbf{x}}}(t) = \mathbf{f}(\hat{\mathbf{x}}(t), \mathbf{u}(t)) + L(t)(\mathbf{y}(t) - \hat{\mathbf{y}}(t)) \tag{9.21}$$

$$\hat{\mathbf{y}}(t) = C\hat{\mathbf{x}}(t) \tag{9.22}$$

where C is the 2×2 identity matrix. Initially, the observer is at the long run equilibrium, so that $\hat{\xi}(0) = [0, 0]^T$. Notice that the observer does not include the expectations variables. These are taken as disturbances on the process, which the observer aims to learn.

The application of the nonlinear observer to this model, is presented as the results of a number of experiments. In the first two experiments the expectations variables are assumed to be zero. The third experiment considers a shock to these expectations variables as they have 'jumped' to some 'news'. The fourth experiment examines the role of the learning gain component.

The state feedback gain matrices, $K(t)$, and the learning gain matrices, $L(t)$ are computed on the basis of linearisation of the nonlinear model, $\mathbf{f}(.)$. The results are presented for Experiments 1 and 2 with this linearisation occurring once, and generating a time invariant observer; and for the linearisation occurring periodically producing the time varying observer. In the Experiments, the linearisation occurs at each time step. With each linearisation, a new state equation driving matrix, A is generated, and from it a new state feedback gain matrix and a new learning gain matrix. Thus the Experiments indicate the benefits of repeated linearisations.

The difference between Experiments 1 and 2 is that the initial linearisation is at a different reference point. For the time invariant observer, the state feedback and learning gains matrices are thus generated from a different but still time invariant state equation driving matrix. These two Experiments indicate the significance of the linearisation point. In Experiment 1, the reference is the long run equilibrium (and observer initial) state ($[0, 0]^T$). This is, probably, the most reasonable linearisation point for it is assumed that the process initial states are unknown. In Experiment 2, the initial linearisation point is the process initial states ($[100, 1]^T$). These choices give a considerable distance between the two linearisation points and ensures that the 'nonlinear region' is included between the two initial linearisation points. Recall that the model is linear in the controls, so that it is only the state linearisation point that is relevant for these Experiments.

In all experiments, the state feedback gain matrix is computed through a Linear Quadratic Regulator. This has the aim of returning the states to their long run equilibrium of zero, and thus removing the business cycle. The weights on the states are $[[2, 0]^T, [0, 2]^T]$, and the weights on the controls are $[[1, 0]^T, [0, 1]^T]$. The learning gain matrix is computed through pole placement, and L is chosen such that the eigenvalues of the learning error driving matrix, $(A - LC)$, are -5 and -2.

The results from Experiments 1 and 2 are shown in Figures 9.5 and 9.6, respectively. The Figures show the results from the two initial linearisation points, and, in each case, the results from a single linearisation or periodic linearisations. The important point the Figures show is that, under these scenarios, the nonlinear observer generates control policies that successfully returns the model to its long run equilibrium. They also show the observer learning in the 'nonlinear region' of the process.

Figure 9.5 shows that, with the model linearised around the long run equilibrium for the observer, GDP is quickly returned to its long run equilibrium value. This is faster than in the case of interest rates, despite GDP being much larger in magnitude. Comparing the time varying and time invariant observers, it can be seen from the Figure, that the time varying observer performs worse than the time invariant observer for GDP. Using the 2-norm as a metric for observer performance, as it, in this case, is the distance from the long run equilibrium, then, for GDP, the time varying observer has a metric of 407 compared to 357 for the time invariant observer. For interest rate, there is little difference between the performance of the two observers. The performance metrics are 11 for the time invariant, and 10 for the time varying observers. For Experiment 1, there is little benefit from periodic linearisation.

The results for Experiment 2 are shown in Figure 9.6, where, again, the time trajectories of the outputs are plotted. In this Experiment, the linearisation point is $[100, 1]^T$. The results for GDP are similar to Experiment 1, with the control policies quickly moving GDP to its long run equilibrium. Again, the time invariant observer performed better than the time varying observer. The metrics are 379 and 408 for the time invariant and time varying observers, respectively. The results for interest rates are more stark. In this case, the time varying observer performs much worse than the time invariant observer. The respective metrics are 10 for the time varying and 7 for the time invariant. For interest rates, the initial linearisation point makes much more difference than for GDP.

Figures 9.5 and 9.6 show that the nonlinear observers are successful with this highly nonlinear model. They also show that there is little benefit to be gained by periodic linearisation. In these Experiments, the time varying observers did not perform as well as the time invariant observers. Experiment 4 also considers this issue.

Further insight into the time varying observer can be obtained from Figure 9.10. In that Figure, using Experiment 1, the elements of the learning gain matrices and the state feedback gain matrices are plotted against time for the early part of the time horizon. For these results, the gain matrices were generated at each time step. The time invariant gains are those for the first time period. The plots show that the elements quickly converge to a constant value. Two of the elements in the state feedback gain matrix change significantly during the initial time periods, indicating that during this period re-linearisation is beneficial. After this period, there may be little benefit by re-linearisation and computation of gain matrices. This may explain why the time varying observers did not perform significantly better than the time invariant observers over the entire time horizon.

It is interesting to compare the control policies generated by the two Experiments. These are plotted in Figures 9.7 and 9.8. In Experiment 1, as can be seen from Figure 9.7, the fiscal control policies differ little between the time varying and time invariant observers. But for monetary policy there is a substantial difference, with the maximum control for the time invariant observer being 0.02, compared to 43 for the time varying observer.

For Experiment 2, as can be seen in Figure 9.8, the fiscal control for the time varying

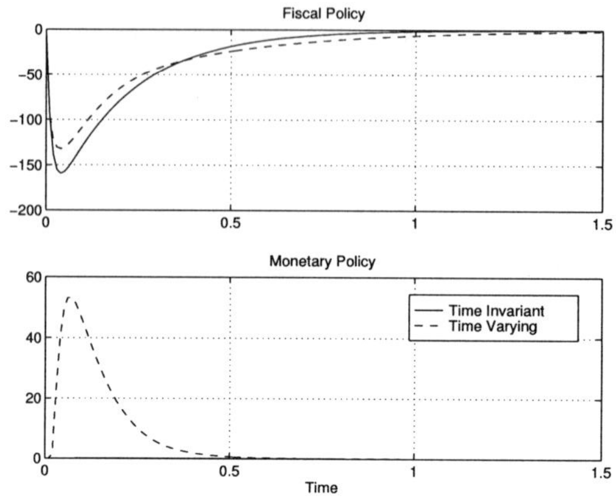

Figure 9.7: Controls: Experiment 1

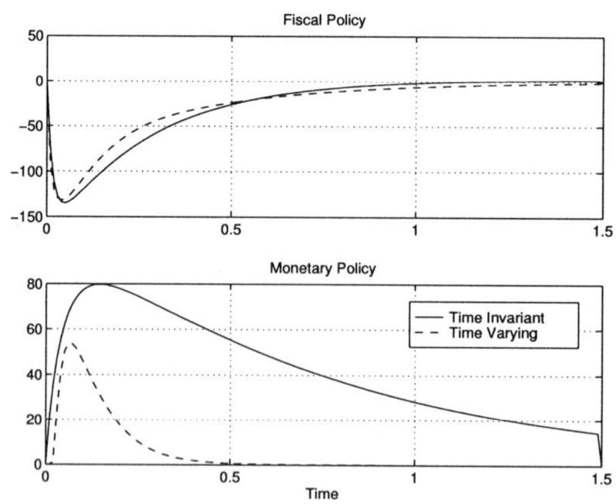

Figure 9.8: Controls: Experiment 2

and time invariant observers are very similar. They also follow the same pattern as in Experiment 1 (Figure 9.7), but are slightly weaker. Again, in Experiment 2, there is a substantial difference between the time varying observer and time invariant observer with monetary policy, but in this case it is in the opposite direction. In Experiment 1, the monetary policy was weak for the time invariant observer, but in Experiment 2 it was strong. On the other hand, there was very little change in monetary policy for the time varying observer between these two experiments.

These results show that there is very little difference in fiscal policy controls generated by the different observers and different initial linearisation points. This is also the case for the two different initial linearisation points with the time varying observer for monetary policy. But with the time invariant observer, the initial linearisation point generates a significant difference for monetary policy.

These results support a continual theme of this book. That is, that many simulation experiments of many and varied conditions are necessary in a search for common policies. This is enhanced by using a wide variety of models. Control techniques are powerful tools that can help in economic policy analysis, but a wide variety of experiments are necessary. Thus to be really useful these tools have to be available and usable by practical economists. This book suggests that visual simulation is a useful approach for this purpose.

Experiment 3 considers a disturbance to the process but not to the observer. The model contains exogenous expectations of forward looking economic actors. The first two Experiments have shown that the unknown initial states of the model can be learned. While not directly included in this model, if the states had been expectations, formed, for example, by the rational expectations hypothesis and had unknown initial values, then those states, too, could have been learned and used in the development of control policies. Again, while not directly showing it here, if the observer knew completely the exogenous shocks, then it too could incorporate these. Experiment 3 considers the question: 'can the observer learn about and develop control policies in a situation where there is an exogenous shock to the process but not the observer?' The shock, in this case, is a change in the expectations of

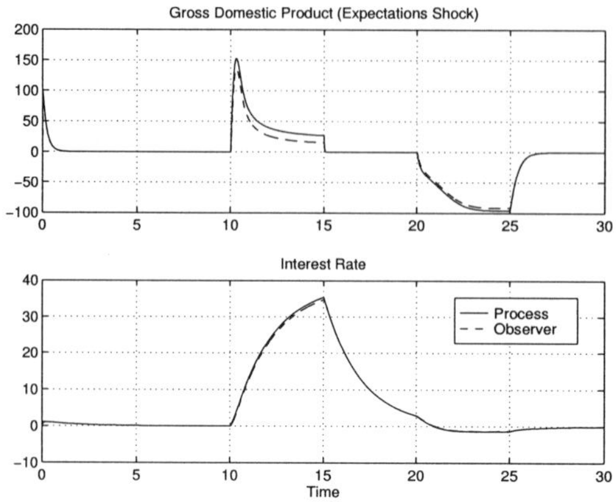

Figure 9.9: Outputs: Expectations Shock

economic actors.

The process has a state equation given by Equation 9.20. All states are considered to be outputs. The observer has the state equation given by Equation 9.21, which does not include the exogenous variables. In Experiments 2 and 3, the exogenous variables where zero for all time, so this difference had no effect. In Experiment 3 the exogenous variables are shock. The first shock is from time 10 to 15, and is a positive shock of the same values as the initial state values, $[100, 1]^T$. The second shock is from time 20 to 25 and is a negative shock of one fifth of the initial values, that is $[-20, -0.2]^T$. The first shock implies that economic actors, due to some 'news', suddenly expect GDP and interest rates to rise substantially. Similarly, the second shock implies some opposite 'news'.

This Experiment uses a time invariant observer under the same scenario as in Experiment 1. The results are plotted in Figure 9.9, and show that the observer can learn about the shocks and generate control policies that take them into account. Recall that the model is highly nonlinear and very sensitive to parameter variations, so that a small shock substantially effects the model. This nonlinearity can be seen by the different effects of negative and positive disturbances on the model responses. Despite the observer not incorporating these

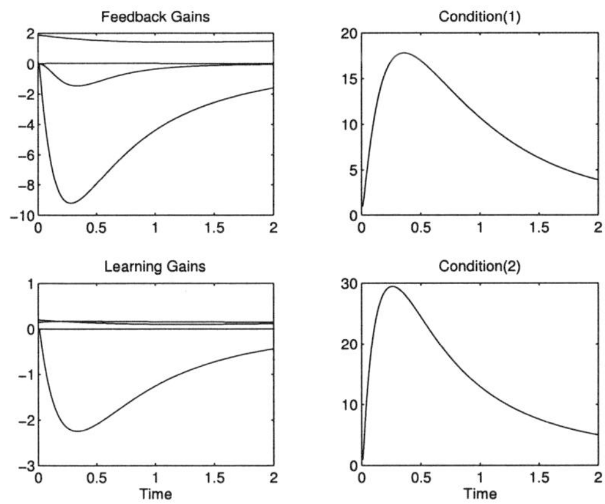

Figure 9.10: Gains and Condition Numbers: Experiment 1

disturbances, it can learn about them and still generate control policies for the model.

During the period of the shock, there is a steady state error between the observer and the process. As the control policies are generated on the observer, and it is not constrained to making economic sense, then some extension to it, by, say, adding some integral action, could be used to overcome the steady state error. That is, the observer could be extended to incorporate greater learning for exogenous shocks.

It has been argued that one way of considering the implications of computing the learning and feedback gain matrices by linearising the nonlinear model, is to consider the linear system matrices, and hence the resulting gain matrices, as not known with certainty. Further, it has been argued that an appropriate metric to measure this uncertainty is the condition number of the eigenvector matrix associated with the closed loop driving matrix. The selection of gains is about the placement of eigenvalues, so how these are effected by perturbation is a metric for the implications of uncertainty. The appropriate driving matrices are the closed loop state equation driving matrix, $[A(t) - B(t)K(t)]$, and the closed loop learning error driving matrix, $[A(t) - L(t)C(t)]$. These are the time varying analogues of the matrices in Equation 9.16. The condition number of the associated eigenvector ma-

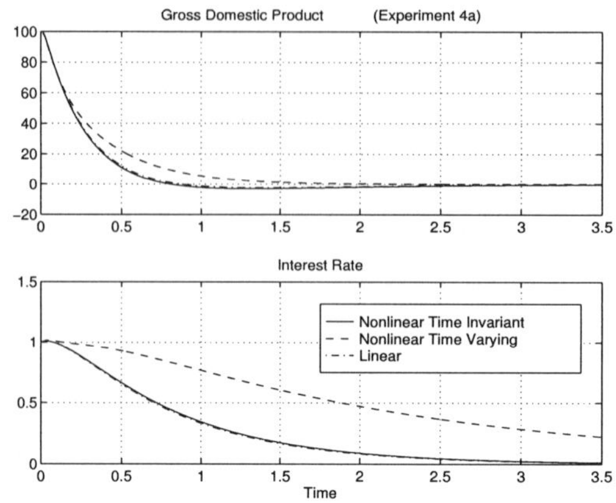

Figure 9.11: Experiment 4: Linear & Nonlinear Observers

trix, is a metric for the effects of a perturbation on the eigenvalues of these matrices. An appropriate condition number is that using the 2-norm for it is the ratio of the largest to smallest singular value. Using Experiment 1, the condition numbers are plotted for the early part of the time horizon in Figure 9.10. The graph labelled 'Condition(1)' is for the closed loop state equation driving matrix, and that labelled 'Condition(2)' is for the closed loop learning error driving matrix. The graphs show that the metric quickly converges. This is as expected, for the elements of the matrices quickly converge. More importantly, they indicate that the associated eigenvector matrices are well-conditioned. Under this scenario and using this metric, uncertainty is not a major problem. Small perturbations to the linear system parameter matrices, via, say, different linearisation points, or uncertainty in the parameters of the original model, will not substantially affect the eigenvalues of the important closed loop matrices.

Experiment 4 considers the role of the learning component in the observer. The nonlinear observer approach of this Chapter has been to extend the conventional (as in Chapter 2) linear observer into nonlinear models. The essential idea is to replace the linear basis model in the observer with the nonlinear model, and to compute the learning and state

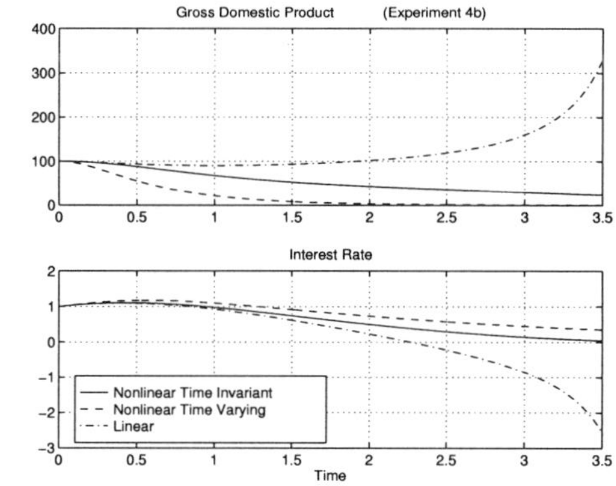

Figure 9.12: Experiment 4: Linear Observer Failing

feedback gains on a linearised version of the nonlinear model. Periodic linearisations may be beneficial.

In earlier Chapters, linear observers were used with nonlinear models. These relied upon the use of a forceful learning term to learn about the effects of the nonlinearities (as well as other effects). This raises the question: 'can a nonlinear observer perform in a situation where a linear observer cannot?' Further, can a time varying nonlinear observer perform in situations where a time invariant nonlinear observer cannot?

Experiment 4 considers these questions using the same scenario as in Experiment 2. It shows that if the learning component is strong enough, then a conventional linear observer can generate policies that return the model to its long run equilibrium. That is, the effects of the nonlinearities can be learned by the linear observer. This is shown in Figure 9.11. The Figure plots the outputs of the process (i.e. Equations 9.18 and 9.19) under the same scenario as in Experiment 2, using a nonlinear time varying, nonlinear time invariant, and linear observers. The Figure shows that all three observers can develop control policies that return the process to its long run equilibrium. The Figure also shows that the observer formed by the linear model performs as well as the nonlinear observers.

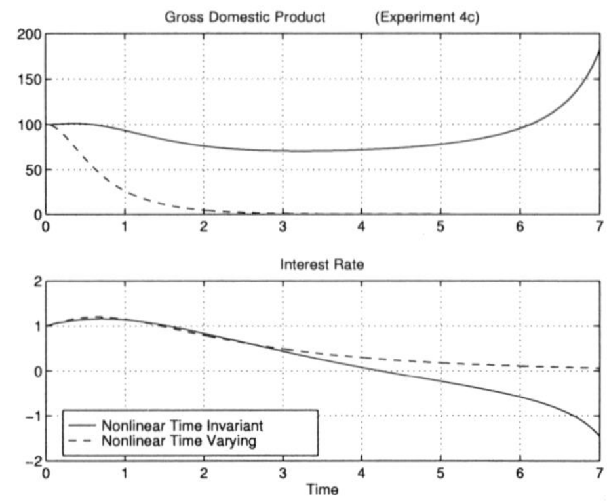

Figure 9.13: Experiment 4: Time Invariant Observer Failing

Figure 9.12 plots the outputs under the same scenario but with a less forceful learning term. In this case the learning gain matrix was computed so that the eigenvalues of the learning error driving matrix are -0.1 and -0.04. The Figure shows that the linear observer is starting to explode. It fails to return the process to its long run equilibrium. It cannot learn about the effects of the nonlinearities in the model. The two nonlinear observers, though, can still control the model with the less forceful learning component. In this situation a nonlinear observer is successful where a linear observer is not.

Figure 9.13 plots the outputs of the process under the two nonlinear observers with an even weaker learning component. In this case, the eigenvalues of the closed loop learning error driving matrix are are -0.05 and -0.02. That is, they are half the magnitude of those used in Figure 9.12. The Figure shows that the time varying observer can control the process whereas the time invariant observer cannot. The benefits of periodic linearisation outweigh the weaker learning component.

A conclusion from Experiments 1 and 2 was that there was little benefit to periodic linearisation, for the time invariant nonlinear observer performed better than the time varying nonlinear observer. This is not the case in Experiment 4 under the weaker learning term. This indicates the trade-off in observer design between the learning component learning about the nonlinearities, and the periodic linearisation account for the nonlinearities in the model.

Experiment 4 has given a specific situation of scenarios when nonlinear observers can work and a linear observer cannot work. It has also shown that this can be extended to the situation of a time varying nonlinear observer working when a time invariant nonlinear observer cannot. The specific situations were generated through the strength of the input-injection learning term. The Experiment highlights the benefits of using nonlinear observers. It also highlights the importance of the learning component, for, it shows, that under the right conditions a linear observer can learn about the effects of the nonlinearities in the process for which it is generating control policies.

9.5 Macroeconometric Model Application

This final nonlinear observer Section applies the nonlinear observer to the practical macro-econometric model constructed in Chapter 7. The time invariant nonlinear observer is illustrated. This is the nonlinear observer where feedback gain matrices are determined on a single linearisation of the model. The basis model for the nonlinear observer is the same Lotka-Volterra IS-LM Model, of Equations 9.18 and 9.19, used in the previous Section. The aim of this illustration is to push the nonlinear observer considerably, to see the power of the output-injection learning term. This final illustration also combines many of the concepts and tools of this book.

For this illustration, the basis model of the observer and the economic system or process, which it is learning about and for which it is developing control policies, are significantly different. The observer basis model has two states and is a low order model, whereas the practical macroeconometric model is a high order model with 258 states. The basis model is highly nonlinear, sensitive to parameter variation, and was designed to be illustrative of the nonlinear observer techniques. This compares to the practical macroeconometric model which was designed to be a realistic model of the Australian economy. The basis model is in

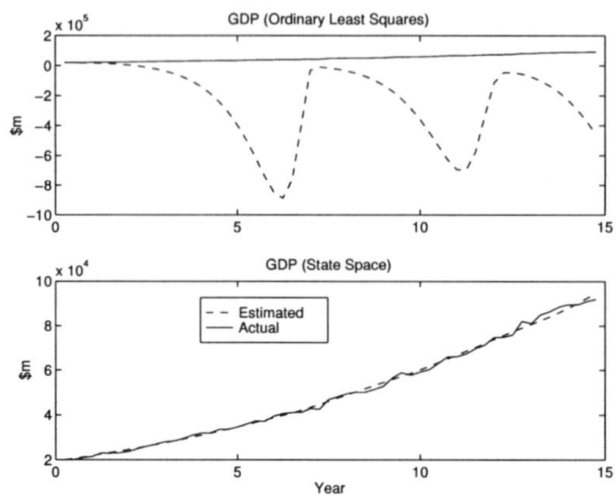

Figure 9.14: Nonlinear Model: GDP Parameters

deviations about an equilibrium, whereas the macroeconometric model is in levels. Finally, the basis model is a continuous time model, but the macroeconometric model is in discrete time. Despite all these differences, the illustration will show that the nonlinear observer can be used to learn about, and develop control policies for the practical macroeconometric model.

The first issue concerns a set of parameters for the basis model. The data used in this illustration is the Historical Baseline data. This is the endogenous variables produced by the practical macroeconometric model where the exogenous variables uses historical data. As with the construction of other econometric models, two methods of parameter estimation were used. The model is linear in its parameters, so the parameter estimation technique of ordinary least squares can be used. The nonlinear least squares, state space estimation procedure is also used. This parameter estimation technique emphasises the output of the model when run out from initial conditions. In this later case, the 2-norm of the estimation error for the interest rate equation (Equation 9.19) was weighted at 1000 times that of GDP (Equation 9.18), due to the difference in magnitude of the data series.

The outputs from both models, when run out from initial conditions, and the Historical

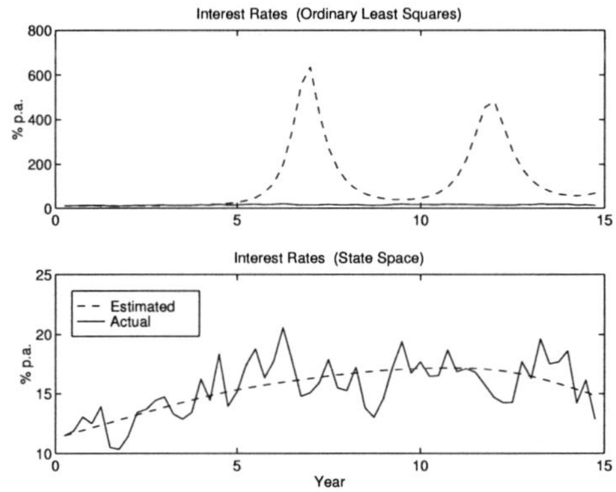

Figure 9.15: Nonlinear Model: Interest Rate Parameters

Baseline data are plotted in Figure 9.14 and 9.15. Figure 9.14 plots the trajectory of GDP. The upper portion of the Figure uses the ordinary least squares estimated parameters of GDP, and the lower portion uses the state space estimated parameters. The superior fit, when the model is run out from initial conditions, of the state space estimation procedure can be seen from the Figure. Figure 9.15 is the analogous trajectories for interest rates. Again, the superiority of the state space estimation procedure can be clearly seen from the Figure.

The visual simulation environment for the basis model with fixed parameter values is presented in Figure 9.16. This Figure is directly comparable to Figure 9.2. The differences are that parameter value objects (e.g. *beta_1*) are constant values and no longer slider-bar windows; the inputs to the object are vectors, which are split into individual data flows; and, similarly, the individual variable data flow outputs are joined to form vectors. A more major difference, though, is that the output-injection learning term has been added to the model. The Figure shows, that visually, this is a simple enhancement to the model, with the *sum* objects being expanded to include an additional variable data flow.

The overall visual simulation environment is presented in Figure 9.17. This is the same overall structure as used throughout this book. It is the same visual simulation environment

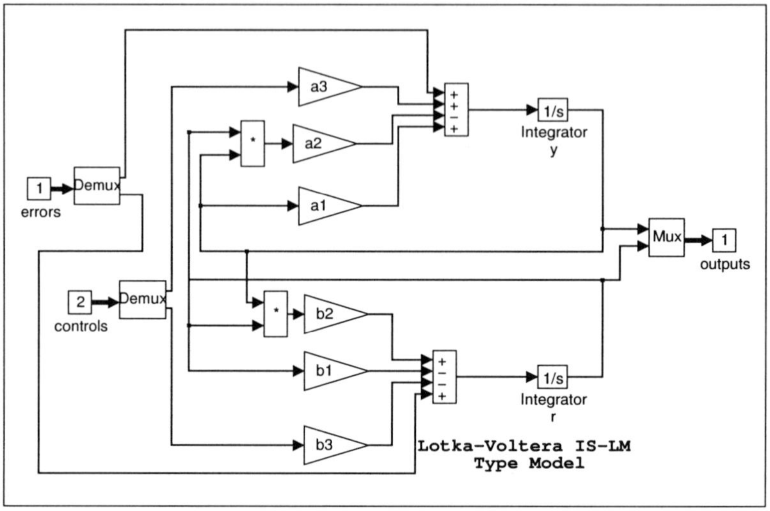

Figure 9.16: *Model* Object

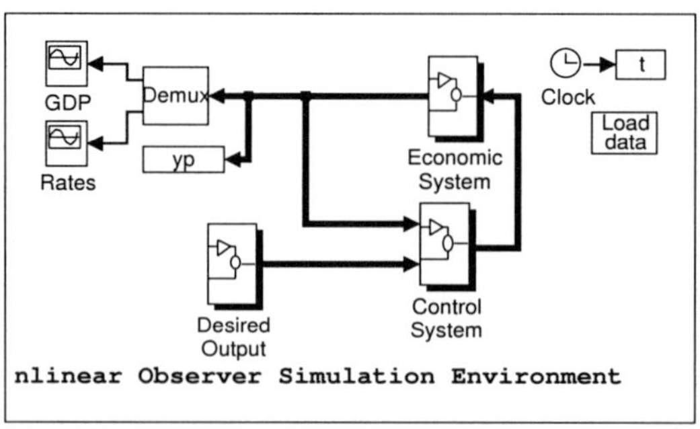

Figure 9.17: Nonlinear Observer Simulation Environment

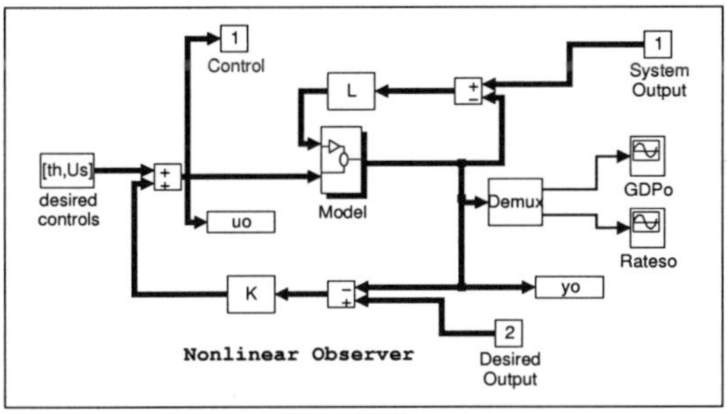

Figure 9.18: *Control System* Object

as given in Figure 8.13 (on page 231) except for the objects used to plot the output. The *Economic System* object is the practical macroeconometric model. It can be seen visually in Figure 8.14 (on page 232). The *Desired Output* object is also similar to those used previously. The scenario for this illustration is that policy makers desire that GDP follows the Historical Baseline data, and that interest rates remain fixed at a constant value. This is formed through a *look-up table* object.

It is the *Control System* object that differs in this illustration. A different control methodology for tracking the desired output is developed for the illustration. Again, the visual simulation tools allow this to be viewed as an extension of previous techniques, and so highlights the benefits of the visual simulation environment. The basic approach of the control technique is to 'invert' the model to obtain a set of controls that will force the model to track the desired output. This is essentially the same idea as was used in the Feedforward method developed in Chapter 4. Using this set of controls as the desired controls, then state feedback is used to force the difference between the model's output and the desired output to zero.

The *Control System* object is presented in Figure 9.18. The state feedback loop around the *Model* can be clearly seen, with the difference between the *Desired Output*, and the model's output flowing into the state feedback gain matrix object, K. When this difference is zero, then the *desired controls* data flow will ensure the model output is the desired output.

From Figure 9.18, the output-injection learning error feedback loop can also be clearly seen. The error between the *Model* output and the *Economic system* output, flow through the learning error gain matrix object, L, and into the *Model*. The Figure also shows the control flowing out of the *Control System*. Display objects are also included in the Figure.

The nonlinear observer used in this illustration is the time-invariant one with the feedback being computed on a single linearisation of the basis model. The linearisation point was selected to be the mean of the Baseline Historical data.

The basis model for the observer is constructed in continuous time, yet the macroeconometric model is a discrete model with a sample time of one Quarter (of a year). So the linearised version of the model was transformed into a discrete equivalent for that sample time. The state feedback gain matrix was computed with a discrete linear quadratic regulator. The weighting matrices used in the result presented below are

$$Q = \begin{bmatrix} 0.002 & 0 \\ 0 & 2 \end{bmatrix} \qquad R = \begin{bmatrix} 0.0001 & 0 \\ 0 & 0.00001 \end{bmatrix} \tag{9.23}$$

The learning error gain feedback matrix was simply allocated values for the results presented in this illustration. The learning gain matrix is

$$L = \begin{bmatrix} 0.005 & 0 \\ 0 & 0.05 \end{bmatrix} \tag{9.24}$$

It can be seen, though, from Figure 9.15, that the trajectory for interest rates is noisy, indicating that a Kalman filter may be appropriate in this situation.

The results of this experiment for GDP are presented in Figure 9.19. The Figure plots the Historical Baseline trajectory and the output from the practical macroeconometric model. The Figure shows that control system generates control policies such that there is a marginal difference to the Historical Baseline.

Figure 9.20 plots the results for the interest rate output. The weighting matrices are such that this output has been given preference, and this is reflected in the results. This Figure

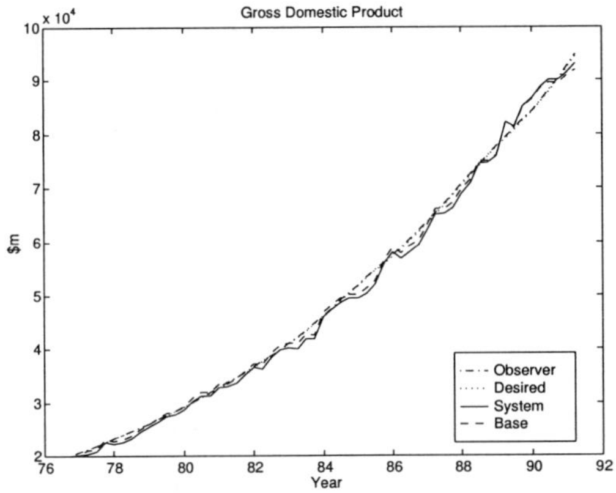

Figure 9.19: Nonlinear Observer Results: GDP

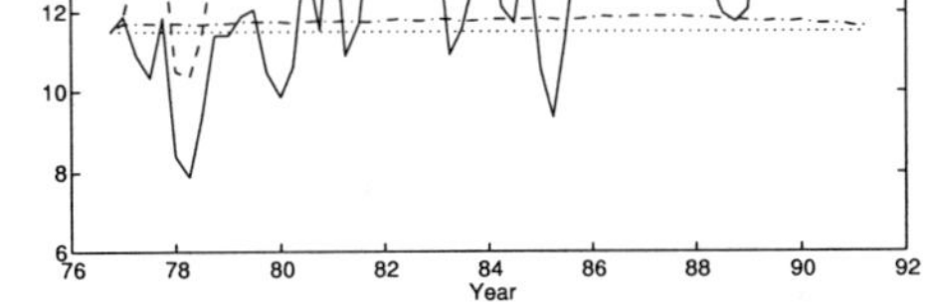

Figure 9.20: Nonlinear Observer Results: Interest Rates

shows that the control scenario lead to an improvement for this output over the Historical Baseline. Using the 2-norm as a metric, the relative error between the Historical Baseline and the desired output is 0.40. Under the control scenario the metric is 0.22.

These results show that the nonlinear observer can be used to generate a set of control policies that can be used in conjunction with the practical macroeconometric model. These control policies were such that the macroeconometric model was closer to a set of desired output trajectories than was the case in the no control situation.

The results show that a low order nonlinear observer that was structurally designed on a completely different basis to a high order practical macroeconometric model, can learn about the effects of these differences and still be used to generate control policy that improves the performance of the *Economic System.* These results show the power of the output-injection learning approach.

9.6 Conclusion

This Chapter has considered a problem that is the focus of much current research in control engineering. This is the problem of the construction of observers for nonlinear systems. The Chapter has developed an approach for nonlinear observers by extending the conventional linear observer. The approach uses a nonlinear model for the basis model in the observer and adds to it an output-injection learning term. The feedback gain matrices, both for the state feedback and the leaning gain feedback, are computed on the basis of linearisation of the nonlinear model. Periodic linearisation, which produces a time varying nonlinear observer, may improve the observer performance.

The nonlinear observer was illustrated with a low order economic model. The model was highly nonlinear and very sensitive to parameter variations. It was based on the Lotka-Volterra equations and formed an IS-LM adjustment type model. The model describes the trajectory of the economy when disturbed from its long run equilibrium.

This Chapter has shown that nonlinear observers can learn about exogenous disturbances as well as the unknown initial values of the states. The unknown initial values of the states may be the values of expectations variables of forward looking economic actors which are generated by the rational expectations hypothesis. The unknown exogenous disturbances may be expectations variables that have 'jumped' to 'news'. These variables are not formed by the rational expectations hypothesis. The nonlinear observer can learn about the expectations of forward looking economic actors.

This Chapter has shown that the nonlinear observer approach can be used to learn the states of the model, and develop control policies that control the model. It has also shown that a nonlinear observer can work in situations where a linear observer cannot. Further, it has shown that a time varying nonlinear observer can work in situations where a time invariant nonlinear observer cannot.

This Chapter has further extended the concept of a nonlinear observer into the situation where the basis model for the observer differs significantly from that of the process it is attempting to control. It used the illustration of the Lotka-Volterra IS-LM model to develop control policies for the high order practical macroeconometric model constructed in this book. It showed that a nonlinear observer worked in this situation.

This Chapter has extended many of the themes of this thesis. Significantly, it has shown that the approach of using small models to learn about unknown effects of larger, more sophisticated models, extends into the nonlinear arena.

Chapter 10

CONCLUSION AND FURTHER WORK

10.1 Synopsis

This book has reported the results of an interdisciplinary study into the analysis of economic policy through the use of control theory techniques. The research has focused, not only on the analysis of economic policy, but also on the development of tools so that the control theory techniques can be presented to, and used by, practising economists. These tools have been model building tools and software tools. This work has combined aspects of control theory, economics and computing.

The major focus of this research has been the analysis of economic systems through the development of optimal policy trajectories. The primary question has been: 'given a macroeconomic system, is it possible to develop, using control theory techniques, a set of policies that move the economic system along a desired time path?' From this follows another important policy question: 'what are the implications and trade-offs between key economic indicators of attempting to achieve such a policy trajectory? These two questions allow for significant examination of the macroeconomic system.

An aim of this book was to show that standard linear quadratic optimal control techniques could be used to develop control policies for linear rational expectations models. This was shown in Chapter 2. In that Chapter, the theoretical framework of output-injection learning was developed. It was shown that, using the rational expectations hypothesis, an observer could be formed that learnt the effects of the expectations variables. The Chapter also showed that the output-injection learning is a separate problem from that of determining a state feedback rule for the model. Thus it is possible to use standard linear quadratic optimal control techniques with linear rational expectations models.

Chapter 2 took as an illustration a model that had been used previously, but unsuccessfully, in an attempt to generate a time consistent optimal control rule. It was shown, using the theoretical framework of the Chapter, that such a rule could be generated.

Another aim of this book was to use best practise econometric models, and to show that best practice control theory techniques could be used in conjunction with these models, without the need to modify the econometric model. That is, the aim was not to use specially constructed econometric models or to ignore aspects of the model. As was shown in Chapter 6, this differs from most applications of control theory to economics. The book has shown that by using separate model based control and output-injection learning, this aim can be achieved.

The use of a separate model based control led to the aim of developing a range of econometric models. Chapter 3 constructed a variety of low order linear dynamic models. In Chapter 9, a low order nonlinear model is constructed. Chapter 3 developed an overall state space framework for the manipulation of dynamic economic models. The advantage of this form is that the models can be used with a wide variety of algorithms, and solution software. These algorithms include standard control theory algorithms. Chapter 3 also specifically examined the problem of contemporaneous control in discrete time models. It showed how such models could be transformed into the standard state space form.

An aim of this book was to construct a variety of dynamic econometric models that could be used in control experiments. Chapter 3 achieved this aim. The models were estimated using standard econometric techniques, and were subject to a set of econometric tests. An estimation procedure, where the parameters of the model were estimated using the state space form, was also used.

A large econometric model was also constructed for use in this book. This allowed the book to achieve its aim of using a premier econometric model for policy analysis. In Chapter 7, the equations and parameters of a major macroeconometric model, were used to construct the model. A solution methodology was developed for the model, and it was transformed into a form that was convenient for policy analysis using control methods.

Another aim of this book was to develop techniques whereby the effects of the uncertainty in these models could by analysed. The book proposed using the condition number as an appropriate metric.

A further aim of this book was to extend the control theory concept of an observer to, not only learning about the unknown initial values of the states, but to learn about differences in model structure, form, and data. This has been achieved by the use of low order linear observers to develop control policies for a high order nonlinear macroeconometric model. Chapter 8 achieved this aim using linear models. Chapter 9 extended this to nonlinear observers. In that Chapter, a low order nonlinear model was shown to be able to learn about, and to develop control policies for, a substantially different model. These differences included both order and structural form.

Yet another aim of this book was to extend this concept of the observer from linear to nonlinear systems, and to construct nonlinear observers. The book has achieved this aim in Chapter 9 where a nonlinear observer was constructed and illustrated. The Chapter developed the theory of a nonlinear observer based on extending the linear observer used in earlier Chapters. It examined the theoretical underpinnings of the nonlinear observer, and showed the importance of the truncation error forcing function. This forcing function is dependent upon the linearisation point. It was shown how a time varying nonlinear observer based on periodic linearisations, could reduce the effects of the truncation error forcing function.

Another control theory aim of this book was to develop techniques for multiple input multiple output systems to track a set of ramps. This aim was achieved in Chapter 4, where the servo problem was extended to multiple inputs and multiple outputs with desired outputs of ramps. The Chapter proposed two methods of tracking a vector of ramps. It compared both these methods, and also compared them to the linear quadratic output tracking method. Both methods extended control theory techniques. The integral method proposed a technique of using double integrators implemented as a small state space system. The feedforward method developed a numerical procedure whereby it is possible to compute the steady state growth rate, and used model inversion to help ensure the ramps were tracked.

An important aim of this book was to develop tools and techniques, so that the ideas of this book could be implemented and used by practising economists. The approach to this aim, was to construct the visual simulation software tools. These tools allow for the development of simulation environments where many and varied experimentation can be undertaken. These tools also allow for the visual explanation of some of the complex mathematical analysis. These tools also use best practice computational techniques. These software tools were developed in Chapter 5, and were used extensively in this book.

An aim of this book was to undertake economic policy analysis via a large number of experiments on a large macroeconometric model and using a variety of techniques. The book has achieved this aim.

10.2 Extensions of this Research

Perhaps the most interesting and obvious extension of this research would be to evaluate the use of the ideas presented in this book by practising economists. While many of the ideas have been discussed and shown to both academic and commercial economists, there has been no formal evaluation of, for example, the visual simulation techniques.

Another area that lends itself to use with the ideas of this book, is the extension of the output-injection learning into the learning of the Kendrick (1981) type. That is, the addition of adaptive control techniques with the observers.

This research has not specifically used game theoretic techniques, nor has it specifically used robust control algorithms. These two related areas could be useful in extending the ideas of policy analysis presented in this book.

The nonlinear observer approach developed in this book is a fruitful area for future research. So, too, is the whole area of nonlinear observers. A more rigorous analysis of the properties of the nonlinear observers developed here, and their role in engineering control applications, is an area that could benefit from more research. A specific issue here is that the linearisation of the nonlinear observer was in two separate spaces. It was linearised in the state and control space for the development of feedback gains, and it was linearised over time in its model simulation. Combining both spaces may lead to interesting results.

This book has argued for many and varied control theory experiments for the analysis of economic policy. It has shown how a wide variety of models can be transformed so that such experimentation can be undertaken. More models, especially major macroeconomic models based on different approaches to model building, would be useful for this policy analysis. The construction of such models in the state space form as used in this book, would be a useful extension of this research.

Many of the variables that are exogenous to the Australian economy are endogenous to other economies. This is becoming increasingly so with the move towards globalisation. Multi-economy models would be a useful extension of this research. The implications of policy design in one economy on another economy, is an interesting issue.

This research focused on output trajectories, so that time was the crucial component. It was about the development of economic policy over time. But time is not the only important variable for economic questions. The spatial distribution of economic issues is also an important policy area. For example the distribution of household income over Australia, or the distribution of population within a city are part of economic policy questions. The techniques developed here are equally applicable to these questions, and could be fruitful areas of research. Other time related issues, such as the development of policies to reduce green house gases, are also applicable to the techniques presented in this book. They, too, could be fruitful areas of further research.

An important area of this research was the whole issue of small separate model based control, with the use of output-injection learning. Further analysis of this whole issue would be a useful extension of this book.

BIBLIOGRAPHY

Aitken, V. C. and Schwartz, H. M. (1994), 'On the exponential stability of discrete-time systems with application to observer design', *IEEE Transactions on Automatic Control* **39**(9), 1959–1962.

Amman, H. M. and Kendrick, D. A. (1993), 'Forward looking behaviour and learning in stochastic control', *The International Journal of Supercomputer Applications* **7**(3), 201–211.

Amman, H. M. and Kendrick, D. A. (1997), Should macroeconomic policy makers consider parameter variances and covariances, *in* 'Conference on Computing in Economics and Finance', Society of Computational Economics.

Amman, H. M., Kendrick, D. A. and Neudecker, H. (1996), Numerical steady state solutions for nonlinear dynamic optimisation models, *in* 'Conference on Computing in Economics and Finance', Society of Computational Economics.

Amman, H. M. and Neudecker, H. (1997), 'Numerical solutions for algebraic matrix riccati equation', *Journal of Economic Dynamics and Control* **21**(2–3), 363–369.

Anderson, B. D. (1993), 'Controller design: Moving from theory to practice', *IEEE Control Systems* pp. 16–25.

Anderson, B. O. and Moore, J. (1990), *Optimal Control: Linear Quadratic Methods*, second edn, Prentice-Hall, Sydney.

Aoki, M. (1967), *Optimization of Stochastic Systems*, Academic Press, New York, U.S.A.

Aoki, M. (1990), *State Space Modelling of Time Series*, second edn, Springer-Verlag, Berlin, Germany.

Arthur, W. B. (1990), 'Positive feedbacks in the economy', *Scientific American* pp. 80–85.

Åström, K. J. (1970), *Introduction to Stochastic Control Theory*, Academic Press, London.

Åström, K. J. (1991), Adaptive control, *in* 'ITAC91', ITAC, Singapore.

Åström, K. J. and Wittenmark, B. (1990), *Computer Controlled Systems: Theory and Design*, second edn, Prentice-Hall, Englewood Cliffs, NJ, USA.

Åström, K. J. and Wittenmark, B. (1995), *Adaptive Control*, second edn, Prentice-Hall, Englewood Cliffs, NJ, USA.

Åström, K. J. and Wittenmark, B. (1997), *Computer Controlled Systems: Theory and Design*, third edn, Prentice-Hall, Englewood Cliffs, NJ, USA.

Başar, T. and Bernhard, P. (1991), *H^{∞} Optimal Control and Related Minimax Design Problems*, Birkhäuser, Berlin, Germany.

Beavis, B. and Dobbs, I. M. (1990), *Optimisation and Stability Theory for Economic Analysis*, Cambridge University Press, London.

Becker, R., Hall, S. and Rüstem, B. (1994), 'Robust optimal decisions with stochastic nonlinear economic systems', *Journal of Economic Dynamics and Control* **18**, 125–147.

Begg, D. K. H. (1982), *The Rational Expectations Revolution in Macroeconomics*, Philip Allen, London, UK.

Bell, R. D. (1971), 'Simplified variance design for parameter estimates in the presence of coloured noise', *Electronics Letters* **7**(13), 369–371.

Bergstrom, A. R. (1976), *Statistical Inference in Continuous Time Economic Models*, North-Holland Publishing Company, Amsterdam.

Bestle, D. and Zeitz, M. (1983), 'Canonical form observer design for non-linear time-variable systems', *International Journal of Control* **38**(2), 419–431.

Blackburn, K. (1987), 'Macroeconomic policy evaluation and optimal control theory: A critical review of some recent developments', *Journal of Economic Surveys* **1**, 111–148.

Blackmore, P. A. and Bitmead, R. R. (1995), 'Duality between the discrete-time Kalman Filter and LQ control law', *IEEE Transactions on Automatic Control* **40**(8), 1442–1444.

Blanchard, O. J. and Fischer, S. (1989), *Lectures on Macroeconomics*, The MIT Press, Cambridge, Massachusetts.

Blanchard, O. J. and Kahn, C. M. (1980), 'The solution of linear difference models under rational expectations', *Econometrica* **48**(5), 1305–1311.

Boucekkine, R. (1995), 'An alternative methodology for solving nonlinear forward-looking models', *Journal of Economic Dynamics and Control* **19**, 711–734.

Bryant, R. C., Mann, C. and Hooper, P. (1993), Stochastic simulations with simple policy regimes, *in* Bryant and Mann (1993), chapter 5.

Bryant, Ralph C., P. H. and Mann, C., eds (1993), *Evaluating Policy Regimes: New Research in Empirical Macroeconomics*, The Brookings Institution, Washington, DC, USA.

Bryson, A. E. Y.-C. and Ho (1975), *Applied Optimal Control: Optimisation Estimation and Control*, Hemisphere Publishing Corp., New York.

Buhly, H. U. (1985), Some environmental aspects of optimal economic growth, *in* Feichtinger (1985).

Burden, R. L. and Faires, J. D. (1985), *Numerical Analysis*, fifth edn, PWS Publishing Company, Boston, USA.

Camacho, E. F. (1993), 'Constrained generalized predictive control', *IEEE Transactions on Automatic Control* **38**(2), 327–332.

Carraro, C. and Sartore, D., eds (1987), *Developments of Control Theory for Economic Analysis*, Kluwer Academic Press, Boston.

Casares, E., Garaza, G. and Kelly, R. (1989), Stabilization of a closed economy model via adaptive control, *in* Christodoulakis (1989*b*), pp. 95–100.

Challen, D. W. and Hagger, A. J. (1981), *Unemployment and Inflation: An Introduction to Macroeconomics*, Longman Chesire Pty. Ltd., Sydney.

Challen, D. W. and Hagger, A. J. (1983), *Macroeconometric Systems: Construction, Validation and Applications*, The Macmillan Press Ltd., London.

Chen, D. and Moler, C. (1993), *Symbolic Math Toolbox: User's Guide*, The Mathworks Inc, Natick, Massachusetts, USA.

Chiang, R. Y. and Sanfonov, M. G. (1992), *Robust Control Toolbox: User's Guide*, The Mathworks Inc, Natick, Massachusetts, USA.

Chow, G. C. (1981*a*), Estimation and optimal control of dynamic game models under rational expectations, *in* Lucas and Sargent (1981), chapter 34.

Chow, G. C. (1987), Developments of control theory in macroeconomics, *in* Carraro and Sartore (1987).

Chow, G. C. (1992), 'Dynamic optimisation without dynamic programming', *Economic Modelling* .

Chow, Gregory, C. (1975), *Analysis and Control of Dynamic Economic Systems*, John Wiley & Sons, New York, USA.

Chow, Gregory, C. (1981*b*), *Econometric Analysis by Control Methods*, John Wiley & Sons, New York, USA.

Christodoulakis, N. M. (1989*a*), 'Extensions of linearisation to large econometric models', *Computer, Mathematics and Applications* **18**(6/7), 629–642.

Christodoulakis, N. M., Currie, D., Garratt, A., Kemball-Cook, D., Levine, P., Barrell, R., Ireland, J. and Westaway, P. (1993), Policy analysis and model reduction techniques with the Global Economic Model, *in* Bryant and Mann (1993), chapter 11, pp. 535–558.

Christodoulakis, N. M., ed. (1989*b*), *Dynamic Modelling and Control of National Economies, 1989*, Permagon Press, Oxford, UK.

Christodoulakis, N. M., Kemball-Cook, D. and Levine, P. (1993), 'The design of economic policy under model uncertainty', *Computational Economics* **6**, 219–240.

Christodoulakis, N. M., Levine, P. and Gaines, J. (1991), 'Macroeconomic policy design using large econometric rational expectations models: Methodology and application', *Oxford Economic Papers* **43**, 25–58.

Christodoulakis, N. M., Levine, P. and Gaines, J. (1993), Macroeconomic policy design using large econometric rational expectations models: Methodology and application, in *Rules Reputation and Macroeconomic Policy Coordination* (Currie and Levine, 1993*c*), chapter 7, pp. 176–209.

Chyung, D. H. (1994), A tracking controller for a class of smooth nonlinear systems, *in* 'Proceedings of 33rd Conference', IEEE Conference on Decision and Control, pp. 3628–3629.

Ciccarella, G., Mora, M. D. and Germant, A. (1995), 'Asymptotic linearization and stabilization for a class on nonlinear systems', *Journal of Optimization Theory and Applications* **84**(3), 495–507.

Cividini, A. and Siviero, S. (1992), Implementing stochastic optimal control of nonlinear models: a comparison of alternative solution methods, Temi di Discussione 179, Banca D'Italia.

C.K.Chui and Chen, G. (1987), *Kalman Filtering with Real Time Applications*, Springer-Verlag, London.

Clark, P. B., Goodhart, C. A. E. and Huang, H. (1996), Optimal monetary policy rules in a rational expectations model of the phillips curve, *in* 'Conference on Computing in Economics and Finance', Society of Computational Economics.

Clements, D. L. (1984), *An Introduction to Mathematical Models in Economic Dynamics*, Polygonal Publishing House, Passaic, NJ, USA.

Cohen, D. and Michel, P. (1988), 'How should control theory be used to calculate time-consistent government policy?', *Review of Economic Studies* **15**, 263–274.

Crouch, R. L. (1972), *Macroeconomics*, Harcourt, Brace and Jovanovich, New York.

Currie, D. (1985), 'Macroeconomic policy design and control theory – a failed partnership?', *Economic Journal* **95**, 285–306.

Currie, D. and Hall, S. G. (1994), Expectations, learning and empirical macro-economic models, *in* Hall (1994), chapter 6, pp. 89–109.

Currie, D. and Levine, P. (1993*a*), Credibility and time consistency in a stochastic world, in *Rules Reputation and Macroeconomic Policy Coordination* (Currie and Levine, 1993*c*), chapter 5, pp. 122–147.

Currie, D. and Levine, P. (1993*b*), The design of feedback rules in linear stochastic rational expectations models, in *Rules Reputation and Macroeconomic Policy Coordination* (Currie and Levine, 1993*c*), chapter 4, pp. 95–121.

Currie, D. and Levine, P. (1993*c*), *Rules Reputation and Macroeconomic Policy Coordination*, Cambridge University Press, Cambridge, U.K.

Currie, D. and Levine, P. (1993*d*), Should rules be simple?, in *Rules Reputation and Macroeconomic Policy Coordination* (Currie and Levine, 1993*c*), chapter 6, pp. 148–175.

Darouach, M., Zasadzinski, M. and Hayar, M. (1996), 'Reduced-order observer design for descriptor systems with unknown inputs', *IEEE Transactions on Automatic Control* **41**(7), 1068–1072.

Davis, M. and R.B.Vinter (1985), *Stochastic Modelling and Control*, Chapman and Hall, London.

Deissenberg, C. (1985), Robustifying macroeconometric models, *in* Feichtinger (1985).

Diller, J. (1994), An annual model of the new south wales economy, *in* 'Conference of Economists', Economics Society of Australia, Surfers Paradise, Queensland, Australia.

Dixon, P. B. and Parmenter, B. R. (1994), Computable general equilibrium modelling, Working Paper IP-65, Centre of Policy Studies, Monash University, Melbourne, Australia.

Dixon, P., Parmenter, B., Sutton, J. and Vincent, D. (1982), *ORANI: A Multisectoral Model of the Australian Economy*, North-Holland, Amsterdam.

Dorf, R. (1980), *Modern Control Systems*, Addison-Wesley, New York.

Doyle, J. C., Francis, B. A. and Tannenbaum, A. R. (1992), *Feedback Control Theory*, Maxwell Macmillan International, New York.

Doyle, J. C., Glover, K., Khargonekar, P. P. and Francis, B. A. (1989), 'State-space solutions to standard $\mathcal{H}_2$ and $\mathcal{H}_\infty$ control problems', *IEEE Transactions on Automatic Control* **34**(8), 831–846.

Eker, J. and Åström, K. J. (1996), A nonlinear observer for the inverted pendulum, Technical report, Lund Institute of Technology, Lund, Sweden.

El-Tohami, M., Lovass-Nagy, V. and Mukundan, R. (1983), 'On the design of observers for generalised state space systems using singular value decomposition', *International Journal of Control* **38**(3), 673–683.

Fair, R. C. and Taylor, J. B. (1983), 'Solution and maximum likelihood estimation of dynamic linear rational expectations models', *Econometrica* **51**(4), 1169–1185.

Feichtinger, G., ed. (1982), *Optimal Control Theory and Economic Analysis 1*, North-Holland Publishers, Amsterdam.

Feichtinger, G., ed. (1985), *Optimal Control Theory and Economic Analysis 2*, North-Holland Publishers, Amsterdam.

Fisher, P. (1989), On the use of optimal control in macroeconomic modelling, *in* Christodoulakis (1989*b*), pp. 101–108.

Fisher, P. (1992), *Rational Expectations in Macroeconomic Models*, Vol. 26 of *Advanced Studies in Theoretical and Applied Econometrics*, Kluwer Academic Publishers, London, UK.

Fitzgerald, V. W., Johnston, H. N. and Bayes, A. (1973), An interactive computing algorithm for optimal policy selection with non-linear econometric models, *in* 'Third Conference of Economists', Australian Economics Society, Adelaide.

Flaschel, P. (1993), *Macrodynamics: Income Distribution, Effective Demand and Cyclical Growth*, Vol. 15 of *Dynamische Wirtschaftstheorie*, Peter Lang, Frankfurt au Main, Germany.

Frisch, R. (1933), Propagation problems and impulse response problems in dynamic economics, *in* 'Economic Essays in Honour of Gustav Cassel', Allen and Unwin, London.

Gabish, G. and Lorenz, H.-W. (1987), *Business Cycle Theory: A Survey of Methods and Concepts*, Vol. 283 of *Lecture Notes in Economics and Mathematical Systems*, Springer-Verlag, Berlin, Germany.

Gagon, J. E. and Tryon, R. W. (1993), Stochastic behaviour of the world economy under alternative policy regimes, *in* Bryant and Mann (1993), chapter 9, pp. 507–518.

Gandolfo, G. (1981), *Qualitative Analysis and Econometric Estimation of Continuous Time Dynamic Models*, North-Holland Publishing Company, Amsterdam.

Gandolfo, G. (1985), Feedback policy rules in a continuous time macroeconometric model of the Italian economy, *in* Feichtinger (1985).

Gaschler, H. J. (1996), Enhancement of an algorithm for the optimal control of nonlinear stochastic models, *in* 'Conference on Computing in Economics and Finance', Society of Computational Economics.

Goodwin, R. M. (1990), *Chaotic Economic Dynamics*, Clarendon Press, Oxford, UK.

Grace, A., Laub, A. J., Little, J. N. and Thompson, C. M. (1992), *Control Systems Toolbox: User's Guide*, The Mathworks Inc, Natick, Massachusetts, USA.

Granger, C. W. J. and Teräsvirta, T. (1993), *Modelling Nonlinear Economic Relationships*, Oxford University Press, Oxford, U.K.

Green, M. (1995), An introduction to $\mathcal{H}_\infty$ control, Research discussion paper, Cooperative Research Centre for Robust and Adaptive Systems, Australian National University, Canberra, ACT 0200, Australia.

Hall, S. G., ed. (1994), *Applied Economic Forecasting Techniques*, Harvester-Wheatsheaf, London, UK.

Hall, S. G. and Henry, S. (1988), *Macroeconomic Modelling*, Vol. 172 of *Contributions to Economic Analysis*, North-Holland Publishers, Amsterdam, the Netherlands.

Hall, S. J. (1986), 'Time inconsistency and optimal policy formulation in the presence of rational expectations', *Journal of Economic Dynamics and Control* **10**, 323–326.

Hall, S. J. (1993), 'Modelling the sterling effective exchange rate using expectations and learning', *The Manchester School* **61**(3), 270–286.

Hall, S. J. and Stephenson, M. J. (1989), Optimal control of stochastic non-linear models, *in* Christodoulakis (1989*b*), pp. 123–130.

Hall, S. J. and Stephenson, M. J. (1990*a*), 'An algorithm for the solution of stochastic optimal control problems for large nonlinear econometric models', *Journal of Applied Econometrics* **5**, 393–399.

Hall, S. J. and Stephenson, M. J. (1990*b*), Optimal control of stochastic nonlinear models, *in* S. Henry and K. Patterson, eds, 'Economic Modelling at the Bank of England', Chapman and Hall, London.

Harvey, A. (1981), *Time Series Models*, Phillip Allan, New York.

Hendry, D. F. (1993), *Econometrics: Alchemy or Science?*, Blackwell Publishers, Oxford, U.K.

Hénin, P.-Y. (1986), *Macrodynamics: Fluctuations and Growth*, Rutledge & Kegan Paul, New York.

Herbert, J. M., Herbert, R. D., Johnson, G. T. and Arnfield, A. J. (1996), Matlab as a software debugging tool, *in* '1996 Australian MATLAB Conference', University of Melbourne, Melbourne, Australia.

Herbert, R. D. (1993), An application of control theory to macroeconomic systems, Master's thesis, School of Mathematics, Physics, Computing and Electronics; Macquarie University, Sydney 2109, Australia.

Herbert, R. D. (1995*a*), A learning based approach to optimal control with linear forward looking economic systems, Computing Report 95-186C, School of MPCE, Macquarie University.

Herbert, R. D. (1995*b*), A learning based approach to optimal control with linear stochastic rational expectations models, *in* 'IFAC/IFIP/IFORS/SEDC Symposium on Modelling and Control of National and Regional Economies', International Federation of Automatic Control, Queensland, Australia.

Herbert, R. D. (1996*a*), A learning based approach to optimal control with linear stochastic rational expectations models, *in* Vlačić, Nguyen and Ćećez-Kecmanović (1996), pp. 139–144.

Herbert, R. D. (1996*b*), A low order dynamic macroeconometric model, Computing Report 95-186C, School of MPCE, Macquarie University.

Herbert, R. D. (1996*c*), Observer based control with nonlinear macroeconometric models, *in* 'International Conference on Computing in Economics and Finance', Society of Computational Economics, Geneva, Switzerland.

Herbert, R. D. (1997), Visual simulation of a large macroeconometric model, *in* 'Third International Conference on Computing in Economics and Finance', Society of Computational Economics, Stanford University, California, U.S.A.

Herbert, R. D. (1998), Control analysis of a dynamic economic model, Research Report 9801, Faculty of Management, University of Western Sydney - Hawkesbury.

Herbert, R. D. and Bell, R. D. (1994*a*), Generating optimal policies with the Murphy model, *in* '23rd Conference of Economists', Economic Society of Australia, Queensland, Australia.

Herbert, R. D. and Bell, R. D. (1994*b*), Generating optimal policies with the Murphy model, Computing Report 94-160C, School of MPCE, Macquarie University. Paper given at the 23th Conference of Economists, Gold Coast, Queensland, Australia.

Herbert, R. D. and Bell, R. D. (1995*a*), Low-order observer based optimal control with economic systems, *in* 'Control 1995 Preprints of Papers', Vol. 9, Melbourne, Australia, pp. 411–415.

Herbert, R. D. and Bell, R. D. (1995*b*), Visualisation in the simulation and control of economic models, Computing Report C/TR95.01, School of MPCE, Macquarie University. Paper given at the 24th Conference of Economists, Adelaide, Australia.

Herbert, R. D. and Bell, R. D. (1996*a*), The application of observers in the control of macroeconomic systems, *in* J. J. G. Jr., J. B.Cruz and M. Peshkin, eds, 'Preprints of 13th Triennial World Congress of IFAC', International Federation of Automatic Control, San Francisco, pp. 297–302.

Herbert, R. D. and Bell, R. D. (1996*b*), Simulation and control of economic models using simulink, *in* '1996 Australian MATLAB Conference', University of Melbourne, Melbourne, Australia.

Herbert, R. D. and Bell, R. D. (1997*a*), Linear contemporaneous control models, *in* 'Third International Conference on Computing in Economics and Finance', Society of Computational Economics, Stanford University, California, U.S.A.

Herbert, R. D. and Bell, R. D. (1997*b*), 'Visualisation in the simulation and control of economic models', *Computational Economics* . forthcoming.

Holly, S. and Hughes-Hallett, A. (1989), *Optimal Control, Expectations and Uncertainty*, Cambridge University Press, Cambridge, U.K.

Holly, S., Zarrop, M. B. and Rüstem, B., eds (1979), *Optimal Control for Econometric Models: An Approach to Economic Policy Formulation*, St. Martin's Press, New York, USA.

Hughes-Hallett, A. and M.L.Petit (1988), 'Trade-off reversal in macroeconomic policy', *Journal of Economic Dynamics and Control* **12**, 85–91.

Jelali, M. and Schwartz, H. (1995), 'Nonlinear identification of hydraulic servo-drive systems', *IEEE Control Systems* **15**(5), 17–22.

Johansen, S. and Juselius, K. (1991), 'Testing structural hypotheses in a multivariate cointegration analysis of PPP and UIP for UK', *Journal of Econometrics* **53**(1–3), 211–244.

Johnson, S. C. and Moler, C. (1995), Compiling matlab, Technical report, The Math Works Inc., Natick, MA. USA. Updated Paper originally presented at Usenix Conference on Very High Level Languages, October 1994, Santa Fe, NM., USA.

Kailath, T. (1980), *Linear Systems*, Prentice-Hall, Englewood Cliffs, NJ, USA.

Kalman, R. E. (1982), Identifiability and problems of model selection in econometrics, *in* W. Hildenbrand, ed., 'Advances in Econometrics', Cambridge University Press, London, U.K., chapter 6.

Kendrick, D. (1987), Software for economic optimal control models, *in* Carraro and Sartore (1987).

Kendrick, D. (1994), Computational approaches to learning with control theory, *in* D. Belsley, ed., 'Computational Techniques for Econometrics and Economic Analysis', Kluwer Academic Publishers, Amsterdam, the Netherlands, pp. 75–87.

Kendrick, D. A. (1981), *Stochastic Control for Economic Models*, McGraw-Hill, New York, USA.

Kendrick, D. A. (1988), *Feedback: A New Framework for Macroeconomic Policy*, Kluwer Academic Publishers, London, U.K.

Kendrick, D. A. (1990), *Models for Analysing Comparative Advantage*, Kluwer Academic Publishers, London, U.K.

Kendrick, D. A. (1995), 'Ten wishes', *Computational Economics* **8**, 65–80.

Kendrick, D. A. and Amman, H. M. (1992), A user's guide to DUAL: A program for quadratic-linear stochastic control problems, Technical Paper T90-4, University of Texas, Department of Economics, University of Texas, Austin, Texas 78712, USA.

Kendrick, D. A. and Amman, H. M. (1995), Programming languages in economics, Working paper, University of Texas, Department of Economics, University of Texas, Austin, Texas 78712, USA.

Kim, J.-G. (1995), Delayed robust tracking control for a nonlinear system: Robotic manipulator, *in* 'International Conference of System Structure and Control', International Federation of Automatic Control, Nantes, France, pp. 364–369.

Krener, A. J. and Isidori, A. (1983), 'Linearization by output injection and nonlinear observers', *Systems and Control Letters* **3**, 47–52.

Kuo, B. (1975), *Automatic Control Systems*, third edn, Prentice Hall, Englewood Cliffs, New Jersey, USA.

Kwakernaak, H. and Sivan, R. (1972), *Linear Optimal Control Systems*, John Wiley, New York.

Kyland, F. E. and Prescott, E. C. (1977), Rules rather than discretion: The inconsistency of optimal plans, *in* Lucas and Sargent (1981).

León De La Barra S., B. A., El-Khoury, M. and Fernández F., M. A. (1996), 'On undershoot in scalar discrete-time systems', *Automatica* **32**(2), 255–259.

León De La Barra S., B. A. and Fernández F., M. A. (1994), 'Transient properties of type *m* continuous time scalar systems', *Automatica* **30**(9), 1495–1496.

Levine, P. (1989), The sustainability of optimal policies: Analysis and empirical assessment using the London Business School model, *in* Christodoulakis (1989*b*), pp. 131–137.

Levine, P. and Smith, P. (1988), 'The gains from optimal control in a small econometric model of the UK', *Journal of Economic Dynamics and Control* **12**, 13–18.

Lin, W. and Byrnes, C. I. (1995), 'Remarks on linearization of discrete-time autonomous systems and nonlinear observer design', *Systems and Control Letters* **25**, 31–40.

Livesey, D. A. (1971), 'Optimising short-term economic policy', *Economic Journal* **81**, 525–546.

Long, N. and Vousden, N. (1977), Optimal control theorems, *in* Pitchford and Turnovsky (1977).

Lorenz, H.-W. (1989), *Nonlinear Dynamical Economics and Chaotic Motion*, Vol. 334 of *Lecture Notes in Economics and Mathematical Systems*, Springer-Verlag, Berlin, Germany.

Louis, C. (1996), Control applications of the TRYM Model, *in* Vlačić et al. (1996), pp. 167–177.

Lucas, R. E. (1976), 'Econometric policy evaluation: A critique', *Carnegie-Rochester Conference Series on Public Policy* **1**, 19–46. The Phillips Curve and Labour Markets.

Lucas, R. E. and Sargent, T. J., eds (1981), *Rational Expectations and Econometric Practice*, George Allen & Unwin, London, UK.

Luenberger, D. G. (1971), 'An introduction to observers', *IEEE Transactions on Automatic Control* **16**(6), 596–602.

Lütkepohl, H. (1991), *Introduction to Multiple Time Series Analysis*, Springer-Verlag Publishers, Berlin, Germany.

Malliaris, A. G. and Brock, W. A. (1991), *Stochastic Methods in Economics and Finance*, Elsevier Science Publishers, Amsterdam, the Netherlands.

Manning, R. and Bell, R. (1990), Optimal control – dark horse or favourite in the macro stakes, *in* 'Proceedings of19th Conference of Economists', Australian Economics Society.

Mathews, J. H. (1992), *Numerical Methods for Mathematics, Science and Engineering*, second edn, Prentice-Hall International Inc., Englewood Cliffs, N.J., U.S.A.

Mathworks (1992), *MATLAB: Reference Guide*, The Mathworks Inc, Natick, Massachusetts, USA.

Mathworks (1994*a*), *Real-Time Workshop: User's Guide*, The Mathworks Inc, Natick, Massachusetts, USA.

Mathworks (1994*b*), *SIMULINK: User's Guide*, The Mathworks Inc, Natick, Massachusetts, USA.

Mathworks (1995), *MATLAB Compiler: User's Guide*, The Mathworks Inc, Natick, Massachusetts, USA.

McDonald, I. M. (1996), *Macroeconomics*, second edn, John Wiley & Sons, Brisbane.

McGrattan, E. R. (1994), 'A note on computing competitive equilibria in linear models', *Journal of Economic Dynamics and Control* **18**, 149–160.

McKibbin, W. J. (1987), Numerical solutions of rational expectations models with and without strategic behaviour, Research Discussion Paper 8706, Reserve Bank of Australia, Sydney, Australia.

McKibbin, W. J. (1989), 'Time-inconsistent policy: A survey of the issues', *Australian Economic Papers* pp. 167–178.

McKibbin, W. J. (1991), The australian economy from a global perspective, Brookings Discussion Papers in International Economics 86, The Brookings Institution, Washington, D.C. U.S.A.

McKibbin, W. J. (1993), Stochastic simulations of alternative monetary regimes using the McKibbin-Sachs Global (MSG2) Model, *in* Bryant and Mann (1993), chapter 10, pp. 519–534.

McKibbin, W. J. (1996), The MSG2 and G-Cubed models of the Australian economy, *in* Vlačić et al. (1996), pp. 55–68.

McKibbin, W. J. and Sachs, J. D. (1988), 'Comparing the global performance of alternative exchange arrangements', *Journal of International Money and Finance* **7**, 387–410.

McKibbin, W. J. and Sachs, J. D. (1989), Implications of policy rules for the world economy, *in* R. C. Bryant, D. A. Currie, J. A. Frankel, P. R. Mason and R. Portes, eds, 'Macroeconomic Policies in an Interdependent World', International Monetary Fund.

McKibbin, W. J. and Sachs, J. D. (1991), *Global Linkages: Macroeconomic Interdependence and Cooperation in the World Economy*, The Brookings Institution, Washington, DC, USA.

Michalsks, H. and Mayne, D. Q. (1995), 'Moving horizon observers and observer-based control', *IEEE Transactions on Automatic Control* **40**(6), 995–1006.

Moraal, P. E. and Grizzle, J. W. (1995), 'Observer design for nonlinear systems with discrete-time measurements', *IEEE Transactions on Automatic Control* **40**(3), 395–404.

Murphy, C. W. (1988*a*), 'An overview of the Murphy Model', *Australian Economic Papers* pp. 175–199. Supplementary Conference Volume.

Murphy, C. W. (1988*b*), 'Rational expectations in financial markets and the Murphy Model', *Australian Economic Papers* pp. 61–88. Supplementary Conference Volume.

Murphy, C. W. (1990*a*), The macroeconomics of a macroeconometric model, Technical report, Australian National University, Canberra, Australia.

Murphy, C. W. (1990*b*), The Murphy Model of the Australian economy: The model in detail, Technical report, Australian National University, Canberra, Australia.

Murphy, C. W. (1996*a*), MM2, *in* Vlačić et al. (1996), pp. 9–18.

Murphy, C. W. (1996*b*), Optimal control of MM2, *in* Vlačić et al. (1996), pp. 159–166.

Murphy, C. W. and Brennan, P. (1993), *Managing Australia's Business Cycle: Lessons from the Past and Optimal Policies for the Future*, Vol. 6 of *National Focus*, The National Farmers' Federation, Canberra, Australia.

Murphy, C. W. and Brennan, P. (1994), How to be the best, Economic Insight 19, National Farmers' Federation, Canberra, Australia.

Murphy, C. and Wallis, K. (1992), 'Beware hidden assumptions in economic forecasts', *Australian Financial Review* .

Myoken, H. and Kamiyama, S. (1989), A design of robust policy rules for macroeconomic models, *in* Christodoulakis (1989*b*), pp. 187–194.

Nagatani, K. (1981), *Macroeconomic Dynamics*, Cambridge University Press, London.

Neck, R. (1996), Can budgetary policy planning be improved by optimal control methods? a case study for Austria, *in* 'Conference on Computing in Economics and Finance', Society of Computational Economics.

Neck, R. and Karbuz, S. (1996), Optimal budgetary policies for Austria under alternative target values, *in* '13th Triennial World Congress', Vol. L, International Federation of Automatic Control, Pergamon Press, pp. 273–278.

Neck, R. and Matulka, J. (1994), 'Stochastic optimum control of macroeconometric models using the algorithm OPTCON', *European Journal of Operational Research* **73**, 384–405.

Neck, R. and Posch, U. (1982), On the 'optimality' of macroeconomic policies: An application to Austria, *in* Feichtinger (1982).

Ogata, K. (1990), *Modern Control Engineering*, second edn, Prentice Hall, Englewood Cliffs, New Jersey, USA.

Ogata, K. (1995), *Discrete-Time Control Systems*, second edn, Prentice Hall, Englewood Cliffs, New Jersey, USA.

Ogata, K. (1997), *Modern Control Engineering*, third edn, Prentice Hall, Englewood Cliffs, New Jersey, USA.

Pagan, A. (1996), Whatever happened to optimal control of econometric models?, *in* Vlačić et al. (1996), pp. 151–158.

Parkin, M. and Bade, R. (1992), *Macroeconomics*, Prentice Hall, Englewood Cliffs, New Jersey.

Parmenter, B. R. (1982), Inter-industry analysis: The ORANI model of Australia's industrial structure, *in* L. Webb and R. Allan, eds, 'Industrial Economics: Australian Studies', George Allen and Unwin, Sydney.

Parmenter, B. R. (1995), Forecasting and policy analysis with the Monash model, *in* 'Proceedings of the 24th Conference of Economists', Economics Society of Australia.

Parsell, B. F., Powell, A. A. and Wilcoxen, P. J. (1991), 'The effects of fiscal restraint on the Australian economy as projected by the Murphy and MSG2 models: A comparison', *Economic Record* pp. 97–114.

Passino, K. M. (1995), 'Intelligent control for autonomous systems', *IEEE Spectrum* **32**(6), 55–62.

Petit, M. L. (1987), A system theoretic approach to the theory of economic policy, *in* Carraro and Sartore (1987).

Petit, M. L. (1990), *Control Theory and Dynamic Games in Economic Policy*, Cambridge University Press, Cambridge, UK.

Petkovski, D. B. (1989), Robustness of discrete time systems, *in* Christodoulakis (1989*b*), pp. 153–160.

Phillips, A. W. (1954), 'Stabilization policy in a closed economy', *Economic Journal* **54**, 290–323.

Phillips, C. L. and Nagle, H. T. (1984), *Digital Control System Analysis and Design*, Prentice-Hall, New York.

Pindyck, R. S. (1972), 'An application of the linear quadratic tracking problem to economic stabilization policy', *IEEE Transactions on Automatic Control* pp. 287–300.

Pindyck, R. S. (1973), *Optimal Planning for Economic Stabilization*, North-Holland Publishing Company, Amsterdam.

Pindyck, R. S. (1977), 'Optimal economic stabilization policies under decentralised control and conflicting objectives', *IEEE Transactions on Automatic Control* **22**(4).

Pindyck, R. S. and Rubinfield, D. L. (1991), *Econometric Models and Economic Forecasts*, third edn, McGraw-Hill Publishing Company, New York.

Pitchford, J. and Turnovsky, S., eds (1977), *Applications of Control Theory to Economic Analysis*, North-Holland, Amsterdam.

Powell, A. A. and Murphy, C. W. (1995), *Inside a Modern Macroeconometric Model: A Guide to the Murphy Model*, Vol. 428 of *Lecture Notes in Economics and Mathematical Systems*, Springer-Verlag, Berlin, Germany.

Preston, A. (1977), Existence, uniqueness, and stability of linear optimal stabilisation policies, *in* Pitchford and Turnovsky (1977).

Ratkowsky, D. A. (1983), *Nonlinear Regression Modelling*, Marcel Dekker Inc., New York, USA.

Robertson, P. and Silver, D. (1995), 'Visualisation case studies: Closing the loop', *IEEE Computer Graphics and Applications* **15**(4), 18–20.

Rüstem, B. (1981), *Projection Methods in Constrained Optimisation and Applications to Optimal Policy Design*, Vol. 31 of *Lecture Notes in Control and Information Systems*, Springer-Verlag, Berlin.

Rüstem, B. (1989), Optimal, time consistent robust feedback rules under parameter, forecast and behavioural uncertainty, *in* Christodoulakis (1989*b*), pp. 195–202.

Sacco, P. L. (1992), 'Optimal control of input-output systems via demand management', *Economic Letters* **39**, 269–273.

Sachs, J. D. and McKibbin, W. J. (1985), Macroeconomic policies in the OECD and LDC external adjustment, Working Paper 1534, National Bureau of Economic Research.

Sargent, T. J. (1987), *Dynamic Macroeconomic Theory*, Harvard University Press, Cambridge, Mass. USA.

Seierstad, A. and Sydsaeter, K. (1987), *Optimal Control Theory with Economic Applications*, North-Holland, New York.

Shahian, B. and Hassul, M. (1993), *Control Systems Design Using Matlab*, Prentice-Hall, Englewood Cliffs, NJ, USA.

Simon, H. A. (1952), 'On the application of the servomechanism theory in the study production control', *Econometrica* pp. 247–258.

Simon, H. A. (1956), 'Dynamic programming under uncertainty with a quadratic criterion function', *Econometrica* **24**, 74–81.

Slotine, J.-J. E. and Li, W. (1993), *Applied Nonlinear Control*, Prentice-Hall, Englewood Cliffs, NJ, USA.

Smith, J. and Murphy, C. (1994), 'Macroeconomic fluctuations in the Australian economy', *The Economic Record* **70**(209), 133–148.

Stemp, P. J. and Scarth, W. M. (1996), Zero inflation targets: Central bank commitment and fiscal policy outcomes, *in* 'Conference on Computing in Economics and Finance', Society of Computational Economics.

Sydsaeter, K. (1985), Survey of some recent results on deterministic optimal control, *in* Feichtinger (1985).

Tapiero, C. S. and Sulem, A. (1994), 'Computational aspects in applied stochastic control', *Computational Economics* **7**, 109–146.

Taylor, D. G. (1994), 'Nonlinear control of electric machines: An overview', *IEEE Control Systems* **14**(6), 41–51.

Taylor, J. B. (1981), Estimation and control of a macroeconomic model with rational expectations, *in* Lucas and Sargent (1981), chapter 33, pp. 660–680.

Theil, H. (1958), *Economic Forecasts and Policy*, North-Holland Publishing Company, Amsterdam.

Tinbergen, J. (1952), *On the Theory of Economic Policy*, North-Holland Publishing Company, Amsterdam.

Tsui, C.-C. (1996), 'A new design approach to unknown input observers', *IEEE Transactions on Automatic Control* **41**(3), 464–468.

Tucci, M. P. (1996), The nonconvexities problem in adaptive control models: A simple computational solution, *in* 'Conference on Computing in Economics and Finance', Society of Computational Economics.

Turnovsky, S. J. (1977*a*), On the scope of optimal and discretionary policies in the stabilisation of stochastic linear systems, *in* Pitchford and Turnovsky (1977).

Turnovsky, S. J. (1977*b*), Optimal control of linear systems with stochastic coefficients and additive disturbances, *in* Pitchford and Turnovsky (1977).

Turnovsky, S. J. (1995), *Methods of Macroeconomic Dynamics*, The MIT Press, Cambridge, Massachusetts, USA.

Tustin, A. (1953), *The Mechanism of Economic Systems*, Harvard University Press, Harvard, Mass. USA.

Vlačić, L., Nguyen, T. and Ćećez-Kecmanović, D., eds (1996), *Modelling and Control of National and Regional Economies 1995*, Pergamon, Oxford, U.K.

Walcott, B. L., Corless, M. J. and Zak, S. H. (1987), 'Comparative study of non-linear state-observation techniques', *International Journal of Control* **45**(6), 2109–2132.

Wallis, K. F. (1993), 'On macroeconometric policy and macroeconometric models', *Economic Record* **69**(205), 113–130.

Weerasooriya, S. and Phan, D. T. (1995), 'Discrete-time LQG/LTR design and modelling of a disk drive actuator tracking servo system', *IEEE Transactions on Industrial Electronics* **42**(3), 240–247.

Westaway, P. (1989), Does time inconsistency really matter?, *in* Christodoulakis (1989*b*), pp. 145–151.

Whitely, J. D. (1994), *A Course in Macroeconomic Modelling and Forecasting*, Harvester Wheatsheaf, London, UK.

Williamson, D. (1991), *Digital Control and Implementation*, Prentice-Hall, New York.

Wittenmark, B. (1985), Design of digital controllers - the servo problem, *in* S. Tzafestas, ed., 'Applied Digital Control', Elsevier Science Publishers, London, chapter 2, pp. 51–70.

Zarrop, M. B., Holly, S., Rüstem, B., Westcott, J. H. and O'Connell, M. (1979), Control of the LBS econometric model via a control model, *in* Holly, Zarrop and Rüstem (1979), pp. 231–264.

Appendix A

SOFTWARE ENGINEERING

This research project involved a large software engineering effort. More than 150,000 lines of code were written by the author for the work included in this thesis. As the author has considerable commercial software engineering experience, the software engineering component of the project has not been emphasised.

The software engineering of this project was guided by the author's strongly held belief, that, for this project to be any more than an academic exercise, it had to incorporate useful software tools, and these tools had to be capable of being widely used. From this follow the two guiding principles of the software engineering of this project *viz.* platform independence and as powerful tools as possible. These guiding principles had a major effect on this project. For example, the author's original aim was to use parallel computing techniques, both for the modelling and visualisation. But while this may have enhanced some aspects of the project, such as the visualisation of output data, it would have limited the ability of the project being disseminated.

The principle of platform independence lead to the choice of software. All software used for the results presented in the thesis has been executed on both a MICROSOFT WINDOWS and UNIX X-WINDOWS platform. Some of the results in the thesis were generated on the UNIX platform and some on the PC platform, depending if the author was at Macquarie University or at home.

The powerful tools principle lead to the use of the technical computing software of MATLAB and SIMULINK by the Mathworks Inc. The bulk of the software was written in MATLAB. MATLAB's knowledge of matrix manipulation and linear, time invariant control algorithms reduced, enormously, the effort in constructing the software. Most of the code for this project was written in MATLAB version 4.2c (Mathworks, 1992). The SIMULINK code was generated by version 1.3 of SIMULINK (Mathworks, 1994*b*). Many of the functions from the MATLAB *Control Systems Toolbox* (Grace, Laub, Little and Thompson, 1992) were also used extensively. Other MATLAB Toolboxes used in this project include the *Robust Control Systems* (Chiang and Sanfonov, 1992) and the *Symbolic Math* (Chen and Moler, 1993). Towards the end of this project, the MATLAB version 5.0 and SIMULINK version 2.0 software became available and were used for some results.

Much of the parameter estimation software was written by the author. Some statistical analysis and parameter estimation was undertaken on standard econometric software. This software included the econometrics packages SHAZAM by Kenneth J. White, version 7.0 and ECONOMETRIC VIEWS: MICRO TSP FOR WINDOWS by Quantitative Micro Software, version 1.0A.

The nonlinear econometric model used in this project was based upon the Murphy model of Australia (Murphy, 1990*b*). Mr. Chris Murphy of Econtech Pty. Ltd. kindly supplied a BORLAND TURBO PASCAL executable version of his model (a TURBO UNIT), and some BORLAND TURBO PASCAL source code that provided a 'minimum' front-end to this executable code. A number of BORLAND TURBO PASCAL programs were written as part of the initial prototyping of this project. None of the results presented in this thesis use this software.

The nonlinear econometric model constructed in Chapter 7 is written in MATLAB. For increased speed, it was transformed into C using the MATLAB *Compiler* (Mathworks, 1995). On the PC platform this source code was compiled using WATCOM C/C++ version 10.6, and on the UNIX platform CC was used.

A number of other C++ programs were constructed for the project. They used BORLAND C++, MICROSOFT C++, WATCOM C++, CC and GCC.

For information on the software contact the author at: Faculty of Management, University of Western Sydney - Hawkesbury, Locked Bag #1, Richmond 2753, Australia. The email address is r.herbert@uws.edu.au

INDEX

augmentation, 115, 127, 165

certainty equivalence, xiii, 8, 19
condition number, xiii, 12, 14, 39, 56, 57, 121, 226, 261
contemporaneous control, xiii, 11, 64, 66, 71, 88, 92, 98, 107, 214
control solution, xiii, 197

detectability, xiii

Feedforward Control method, 132
Feedforward method, 110
Final Value Theorem, 113, 114

game, 7, 12, 24, 182

input-injection, 265
Integral Control Method, 114
Integral Control method, 154, 163, 166
Integral method, 110
IS-LM, 64, 67, 68, 107, 248

learning, 3, 8, 14, 27, 41, 159, 227
Linear Quadratic Output Tracking method, 111, 142
linearisation methods, 175

model consistent, xiii, 16, 198
model reduction, 180
models
- Currie & Levine, 36, 155, 167
- GEM, 180
- IS-LM, 67
- LBS, 183, 190
- low order nonlinear, 249
- LPL, 183
- Multiplier Accelerator, 65, 214
- Murphy, 13, 184, 194
- NIESR, 183
- Six State, 82, 232
- Three State, 75, 163
- TRYM, 188
- VAR, 94

modern control theory, 5, 9
multipliers, 39, 41, 49, 112, 159, 169

observer, 3, 10, 14–17, 27, 28, 30–32, 34–36, 39, 41–46, 48, 52, 55, 56, 58, 59, 64, 66, 74, 75, 80, 159, 162, 181, 213, 215, 227
OLS, 76
Optcon Algorithm, 175
output-injection, xiv, 3, 10, 14, 30, 58, 163, 227, 242, 267

Phillips curve, 37, 71, 72, 107, 184, 194

ramp, 11, 109, 161

rational expectations, xiv, 7, 8, 10, 157, 182, 194

saddlepath property, 21, 45
shocked solution, xiv, 197
stationary, xiv, 78, 99, 103
system type, xiv, 109

time-consistent, xiv, 10, 17, 21, 58, 182, 242

visual simulation, 4, 13, 14, 206, 231, 250
visualisation, 4, 152, 169

Advances in Computational Economics

1. A. Nagurney: *Network Economics.* A Variational Inequality Approach. 1993 ISBN 0-7923-9293-0
2. A.K. Duraiappah: *Global Warming and Economic Development.* A Holistic Approach to International Policy Co-operation and Co-ordination. 1993 ISBN 0-7923-2149-9
3. D.A. Belsley (ed.): *Computational Techniques for Econometrics and Economic Analysis.* 1993 ISBN 0-7923-2356-4
4. W.W. Cooper and A.B. Whinston (eds.): *New Directions in Computational Economics.* 1994 ISBN 0-7923-2539-7
5. M. Gilli (ed.): *Computational Economic Systems.* Models, Methods & Econometrics. 1996 ISBN 0-7923-3869-3
6. H. Amman, B. Rustem, A. Whinston (eds.): *Computational Approaches to Economic Problems.* 1997 ISBN 0-7923-4397-2
7. G. Pauletto: *Computational Solutions of Large-Scale Macroeconometric Models.* 1997 ISBN 0-7923-4656-4
8. R.D. Herbert: *Observers and Macroeconomic Systems.* Computation of Policy Trajectories with Separate Model Based Control. 1998 ISBN 0-7923-8239-0

KLUWER ACADEMIC PUBLISHERS – DORDRECHT / BOSTON / LONDON